SURVIVAL PROJECTS BOOK FOR NO GRID LIVING

A Comprehensive Step-by-Step Guide to DIY Projects for Self-Sufficiency: Survive & Thrive Beyond Civilization

James C. Howard

Copyright © 2024 James C. Howard

All rights reserved. No part of this book

may be reproduced, distributed, or transmitted in any form or by any means, including photocopying, recording, or other electronic or mechanical methods, without the prior written permission of the author, except in the case of brief quotations embodied in critical reviews and certain other non-commercial uses permitted by copyright law.

Disclaimer

The information provided in this book, "No Grid Survival Projects: A Comprehensive Step-by-Step Guide to DIY Projects for Self-Sufficiency, Survive & Thrive Beyond Civilization," authored by James C. Howard, is intended for educational and informational purposes only. The author and publisher are not responsible for any consequences resulting from the use, misuse, or application of the techniques, strategies, or projects outlined in this book.

Readers are advised to exercise caution and use their discretion when implementing any of the survival projects or techniques described herein. It is important to recognize that survival situations can be inherently dangerous, and appropriate precautions should always be taken. Before undertaking any project or activity discussed in this book, readers are encouraged to conduct thorough research, seek professional guidance if necessary, and ensure compliance with relevant laws, regulations, and safety standards.

The author has made every effort to ensure the accuracy and reliability of the information presented in this book at the time of publication. However, due to the evolving nature of survival practices and technologies, readers are encouraged to verify the currency and suitability of the information for their specific circumstances.

By reading this book, readers acknowledge and agree that the author and publisher shall not be liable for any direct, indirect, incidental, special, or consequential damages arising out of the use or inability to use the information provided herein, even if advised of the possibility of such damages.

Remember, survival preparedness requires careful planning, practical skills, and a mindset focused on adaptability and resilience. Stay safe, stay informed, and embrace the spirit of self-sufficiency as you embark on your journey beyond civilization.

About the Book

Welcome to the world of self-sufficiency and resilience! In "No Grid Survival Projects: A Comprehensive Step-by-Step Guide to DIY Projects for Self-Sufficiency, Survive & Thrive Beyond Civilization," I invite you on a journey to unlock the secrets of living off the grid and thriving in any environment.

Drawing upon years of experience and a passion for sustainable living, I've crafted this guide to empower you with practical knowledge and hands-on projects that will enable you to embrace a life of independence and preparedness. Whether you're seeking to disconnect from modern utilities or preparing for unforeseen emergencies, this book equips you with the essential skills and insights to navigate the challenges of off-grid living with confidence.

From constructing your own renewable energy systems to cultivating your own food sources and creating sustainable shelter, each chapter is meticulously crafted to provide clear, step-by-step instructions that cater to both beginners and seasoned survivalists alike. Through a combination of proven techniques and innovative approaches, you'll learn how to adapt, innovate, and thrive beyond the constraints of civilization.

But beyond the practical aspects, this book is a celebration of the human spirit and our innate ability to overcome adversity. It's a reminder that self-sufficiency isn't just about surviving—it's about reclaiming our connection to the natural world, fostering resilience, and embracing a simpler, more fulfilling way of life.

So, whether you're dreaming of escaping the confines of urban living or simply looking to enhance your preparedness skills, "No Grid Survival Projects" is your indispensable companion on the journey to self-sufficiency and empowerment.

Join me as we embark on this transformative adventure together.

Warm regards,

James C. Howard

About the Author

Hello, fellow adventurers,

I'm James C. Howard, and I'm thrilled to share with you my passion for self-sufficiency and off-grid living in my book, "No Grid Survival Projects: A Comprehensive Step-by-Step Guide to DIY Projects for Self-Sufficiency, Survive & Thrive Beyond Civilization."

For me, the journey towards self-reliance started with a deep-seated desire to connect with nature and rediscover the lost art of living off the land. Growing up with a love for the outdoors, I've always been fascinated by the idea of crafting my own path and thriving in even the most challenging environments.

Through years of hands-on experience, trial, and error, I've honed my skills in various DIY projects aimed at enhancing self-sufficiency and resilience. From building shelter and purifying water to generating renewable energy and growing food, I've explored it all and learned valuable lessons along the way.

In this book, I've distilled my knowledge, insights, and practical know-how into a comprehensive guide that empowers you to embark on your own off-grid adventure. Whether you're a seasoned survivalist or a curious beginner, my aim is to provide you with the tools, techniques, and inspiration you need to thrive beyond civilization.

But remember, survival isn't just about mastering skills; it's also about cultivating a mindset of adaptability, resourcefulness, and respect for the natural world. As you delve into the pages of this book, I encourage you to embrace the journey, embrace the challenges, and above all, embrace the freedom that comes with self-sufficiency.

Thank you for joining me on this extraordinary odyssey. Together, let's unlock the secrets of off-grid living and discover the boundless possibilities that await us beyond the grid.

Warm regards,

James C. Howard

Contents

Survival Projects book for No Grid Living 1

A Comprehensive Step-by-Step Guide to DIY Projects for Self-Sufficiency: Survive & Thrive Beyond Civilization 1

   James C. Howard 1

  Copyright © 2024 James C. Howard 2

  Disclaimer 3

  About the Book 4

  About the Author 6

   Biography of "No Grid Survival Projects Book" by James C. Howard ... **Error! Bookmark not defined.**

Part 1: 11

Introduction to No Grid Living 11

  Chapter 1: 12

  What is No Grid Living? 12

    Defining Self-Sufficiency 14

    Benefits and Challenges of Off-Grid Life 15

    Different Approaches to No Grid Living 16

  Chapter 2: 18

  Planning for Your Off-Grid Future 18

    Assessing Your Needs, Skills, and Resources 19

    Setting Realistic Goals and Timelines 20

    Choosing a Location and Considering Local Regulations 21

    Financial Considerations for Off-Grid Living 23

Part 2: 25

Building Your Off-Grid Infrastructure 25

  Chapter 3: 26

  Energy Generation 26

    Understanding Your Energy Needs 26

  Chapter 4: 34

  Water Collection and Purification 34

    Water Needs for Drinking, Sanitation, and Irrigation 35

    Greywater Systems 37

    Natural Water Purification Methods 38

  Chapter 5: 40

- Food Production and Preservation 40
    - Planning Your Off-Grid Garden: 40
    - Raising Small Livestock 43
    - Food Preservation Techniques: 45

Part 3: 47

Shelter and Waste Management 47

Chapter 6: 48

Making Your Home More Energy-Efficient 48
- Identifying Energy Leaks in Your Home 50
- Improving Insulation and Air Sealing 51
- Energy-Efficient Appliances 53

Chapter 7: 55

Sustainable Waste Management 55
- Composting Toilets: 56
- Types and Construction Options 56
- Greywater Systems for Non-Drinking Purposes 59

Part 4: 62

Essential Skills and Knowledge for No Grid Living 62

Chapter 8: 63

Tool Maintenance and Repair 63
- Essential Tools for Off-Grid Living 64
- Basic Repairs and Maintenance for Common Tools 65
- Troubleshooting and Problem-Solving Techniques 67

Chapter 9: 69

First Aid and Emergency Preparedness 69
- Building a First-Aid Kit for Off-Grid Situations 71
- Basic First Aid Skills and Response Protocols 72
- Emergency Preparedness Planning 73

Chapter 10: 75

Securing Your Off-Grid Property 75
- Perimeter Security and Deterrents for Trespassers 76
- Securing Your Food and Resources 78
- Off-Grid Communication Options and Emergency Alerts 79

Part 5: 81

Advanced Topics and Considerations 81

Chapter 11: 82
Off-Grid Communication 82
- Ham Radio Communication and Licensing 83
- Satellite Internet Options 84
- Low-Tech Communication Methods 86

Chapter 12: 87
Legal and Regulatory Considerations 87
- Building Permits and Zoning Regulations for Off-Grid Structures 88
- Navigating Zoning Regulations: 89
- Water Rights and Rainwater Harvesting Regulations 90
- Animal Husbandry Regulations 91

Chapter 13: 93
Building a community: Off-Grid Living in Groups 93
- Benefits and Challenges of Community Living 94
- Barter Systems and Local Co-ops 97

Part 6: 99
Appendix 99
- Resource Guide - Books, Websites, and Organizations 100
- Glossary of Terms 103
- Bonus 106
- Additional Resources and Worksheets 106

PART 1

INTRODUCTION TO NO GRID LIVING

CHAPTER 1
WHAT IS NO GRID LIVING?

No Grid Living isn't just a lifestyle choice; it's a philosophy—a profound reimagining of our relationship with the world around us. It's about breaking free from the constraints of modern civilization and forging a deeper connection with nature, reclaiming our autonomy, and embracing a life of self-sufficiency and resilience.

Imagine waking up to the gentle rustle of leaves outside your window, the earthy scent of morning dew hanging in the air. As you step outside, you're greeted by the symphony of nature—the chorus of birdsong, the whisper of the wind through the trees. This is the essence of No Grid Living—a life lived in harmony with the rhythms of the natural world, unburdened by the trappings of modern society.

But what exactly does it mean to live off the grid? At its core, No Grid Living is about disconnecting from centralized infrastructure—electricity, water, sewage—and instead, relying on decentralized, sustainable solutions to meet our basic needs. It's about generating our own power through solar panels or wind turbines, harvesting rainwater for drinking and irrigation, and composting our waste to nourish the soil.

Yet, No Grid Living is more than just a practical solution to environmental challenges—it's a deeply personal journey of self-discovery and empowerment. It's about rediscovering the simple joys of life—the satisfaction of growing your own food, the warmth of a crackling fire on a cold winter's night, the sense of accomplishment that comes from living in harmony with nature.

For our family, No Grid Living was a revelation—a chance to escape the hustle and bustle of city life and reconnect with what truly matters. It was a journey marked by moments of profound beauty and undeniable challenge, a quest for meaning and purpose in a world that often feels disconnected and chaotic.

But amidst the trials and tribulations, they discovered something remarkable—a sense of freedom, of authenticity, of belonging. They found solace in the quietude of the wilderness, strength in the face of adversity, and a deep sense of gratitude for the abundance of the natural world.

So, as we embark on this journey together, let us embrace the spirit of No Grid Living with open hearts and open minds. Let us rediscover the joy of simplicity, the beauty of self-reliance, and the profound wisdom of living in harmony with nature. For in the boundless expanse of the wilderness, there lies a world of possibility—a world where anything is possible, and where the only limit is our imagination.

Defining Self-Sufficiency

Self-sufficiency isn't just a lifestyle choice; it's a philosophy—a deeply ingrained belief in our ability to provide for ourselves and our loved ones, independent of external systems and structures. It's about reclaiming our autonomy, rediscovering the lost art of resilience, and forging a deeper connection with the natural world that sustains us.

Imagine waking up each morning to the gentle caress of a sunbeam streaming through your window, knowing that the warmth it brings is not just a fleeting luxury, but a renewable source of energy that powers your home and fuels your dreams. Picture stepping outside to tend to your garden, feeling the rich earth between your fingers and the cool breeze on your face, knowing that the fruits of your labour will nourish your body and soul for months to come.

But self-sufficiency is more than just growing your own food or generating your own electricity; it's a mindset—a way of life characterized by independence, resourcefulness, and a profound respect for the Earth. It's about learning to live in harmony with nature, to harness its abundance without depleting its resources, and to cultivate resilience in the face of adversity.

For our intrepid family, self-sufficiency became a beacon of hope in a world fraught with uncertainty. It was a journey marked by triumphs and setbacks, by moments of profound connection and undeniable hardship. Yet, through it all, they discovered a profound sense of empowerment—a liberation from the shackles of consumerism, and a newfound appreciation for the simple joys of life.

Self-sufficiency is a journey of self-discovery—a journey that challenges us to question the status quo, to reevaluate our priorities, and to rediscover the inherent wisdom of living in harmony with nature. It's a journey that invites us to embrace the unknown, to step outside our comfort zones, and to chart a course towards a future defined not by scarcity and fear, but by abundance and possibility.

So, as we embark on this journey together, let us embrace the spirit of self-sufficiency with open hearts and open minds. Let us honour the wisdom of our ancestors, who lived in harmony with the land for countless generations. And let us chart a course towards a future where self-sufficiency isn't just a dream, but a reality—a way of life that nourishes our bodies, enriches our souls, and sustains the delicate balance of life on this precious planet.

Benefits and Challenges of Off-Grid Life

Embarking on the journey of off-grid living is a courageous leap into the unknown, a bold declaration of independence from the trappings of modern society. As our intrepid family discovered, this path is paved with both blessings and trials, each one shaping their experience in profound ways.

Benefits:

Freedom and Independence: Perhaps the most profound benefit of off-grid living is the sense of freedom and independence it brings. Freed from the constraints of utility bills and municipal regulations, our family found themselves masters of their own destiny, charting their own course towards a more sustainable and self-reliant way of life.

Connection to Nature: Off-grid living offers a rare opportunity to reconnect with the natural world in a deeply profound way. Surrounded by the majesty of the wilderness, our family found solace in the rhythmic ebb and flow of the seasons, forging a deep and abiding connection to the land that sustained them.

Simplicity and Serenity: In the hustle and bustle of modern life, it's easy to become overwhelmed by the constant barrage of noise and distractions. Off-grid living offers a respite from this chaos, providing a sanctuary of simplicity and serenity where one can find peace amidst the tranquil beauty of nature.

Self-Sufficiency and Resilience: Perhaps the greatest benefit of off-grid living is the sense of self-sufficiency and resilience it fosters. Armed with the knowledge and skills to provide for their own needs, our family found themselves empowered to weather any storm, both literally and metaphorically, with grace and fortitude.

Challenges:

Isolation and Loneliness: While off-grid living offers unparalleled solitude and tranquillity, it can also be isolating at times. Our family often found themselves grappling with feelings of loneliness, yearning for the companionship and camaraderie of their fellow humans.

Resource Limitations: Living off the grid requires a keen understanding of resource management and conservation. Our family quickly learned that every drop of water, every ray of sunlight, and every watt of electricity was precious, and that wastefulness was not an option in their quest for self-sufficiency.

Unpredictable Weather and Environmental Factors: Off-grid living exposes one to the raw power of nature in all its glory—and all its fury. From blistering heatwaves to bone-chilling cold snaps, our family faced a barrage of unpredictable weather and environmental factors that tested their resolve and resourcefulness at every turn.

Technical Challenges: Building and maintaining the infrastructure necessary for off-grid living is no small feat. From installing solar panels to constructing rainwater collection systems, our family encountered a host of technical challenges that required patience, perseverance, and a healthy dose of ingenuity to overcome.

In the end, the benefits of off-grid living far outweighed the challenges for our intrepid family. Despite the trials and tribulations, they faced along the way, they emerged stronger, wiser, and more deeply connected to the land and each other than ever before. And as they looked out over the vast expanse of wilderness stretching out before them, they knew that they had found their true home—a place where they could live in harmony with nature, thrive off the grid, and embrace a life of true freedom and fulfillment.

Different Approaches to No Grid Living

n the realm of self-sufficiency, there exists a tapestry of diverse approaches to living off the grid, each weaving its own unique narrative of resilience, ingenuity, and connection to the land. As our intrepid family set out on their journey towards no grid living, they soon discovered that the path to self-sufficiency was not a one-size-fits-all endeavour but rather a rich tapestry of possibilities, each offering its own set of challenges and rewards.

Minimalist Living: For some, no grid living is synonymous with minimalist living—a return to simplicity, a shedding of excess, and a deepening of one's connection to the essentials of life. Inspired by the principles of minimalism, our family embraced a lifestyle characterized by frugality, mindfulness, and a commitment to living with less. They pared down their possessions to the bare essentials, opting for quality over quantity, and finding liberation in the freedom from material clutter. In the simplicity of their surroundings, they discovered a profound sense of abundance, finding joy in the beauty of the natural world and the richness of human connection.

Homesteading: For others, no grid living is synonymous with homesteading—a return to the land, a revival of traditional skills, and a reclamation of self-sufficiency. Drawn by the call of the wild, our family embraced the homesteading lifestyle, carving out a sanctuary amidst the rolling hills and verdant valleys of the countryside. They cultivated their own food, raised their own livestock, and built their own shelter from the earth beneath their feet. In the rhythm of the seasons, they found solace, finding purpose in the timeless cycle of planting, harvesting, and preserving the bounty of the land.

Permaculture: Yet for others, no grid living is synonymous with permaculture—a holistic approach to sustainable living that seeks to mimic the patterns and resilience of natural ecosystems. Inspired by the wisdom of nature, our family embraced the principles of permaculture, designing their homestead to work in harmony with the earth's natural rhythms. They planted food forests, built swales to capture rainwater, and created living soils teeming with microbial life. In the diversity of their landscapes, they found abundance, finding nourishment in the bounty of perennial crops and the symbiotic relationships between plants, animals, and soil microorganisms.

Urban Homesteading: And let us not forget the urban homesteaders, who bring the principles of self-sufficiency to the heart of the city. In the concrete jungles and asphalt alleyways, they find fertile ground for their dreams of no grid living, transforming vacant lots into thriving gardens, rooftops into solar oases, and community spaces into vibrant hubs of resilience and innovation. Inspired by the pioneering spirit of their rural counterparts, urban homesteaders prove that no grid living is not bound by geography but rather by the boundless creativity and determination of the human spirit.

In the mosaic of no grid living, there exists a multitude of paths to self-sufficiency, each offering its own unique blend of challenges and rewards. Whether minimalist or homesteader, Perma culturist or urban dweller, the journey towards no grid living is a testament to the resilience of the human spirit, a celebration of our innate connection to the land, and a reminder of the endless possibilities that await those bold enough to chart their own course towards a brighter, more sustainable future.

CHAPTER 2
PLANNING FOR YOUR OFF-GRID FUTURE

In the tranquil embrace of nature, our intrepid family found themselves at a crossroads—a pivotal moment in their journey towards self-sufficiency. As they gazed out upon the sprawling expanse of wilderness before them, they realized that success in off-grid living wasn't just about rolling up their sleeves and diving headfirst into the unknown; it was about careful planning, strategic thinking, and a deep understanding of their own needs, skills, and resources.

For our family, planning for their off-grid future began with a candid assessment of their unique circumstances. They took stock of their daily routines, their dietary preferences, and their long-term goals, mapping out a blueprint for their ideal off-grid lifestyle. They asked themselves tough questions, probing the depths of their desires and aspirations, and charted a course towards a future defined by independence, resilience, and boundless possibility.

But planning for their off-grid future wasn't just about envisioning their dream life—it was also about setting realistic goals and timelines. Our family understood that Rome wasn't built in a day, and that their journey towards self-sufficiency would be a gradual, incremental process. So, armed with patience and perseverance, they broke down their goals into manageable milestones, celebrating each small victory along the way as they edged ever closer to their ultimate vision of off-grid paradise.

Choosing a location was another crucial aspect of planning for their off-grid future. Our family understood that not all wildernesses were created equal, and that finding the perfect spot to call home required careful consideration of a myriad of factors—from climate and terrain to access to essential resources like water and sunlight. They scoured maps, conducted site visits, and consulted with local experts, searching for that elusive patch of paradise where their off-grid dreams could take root and flourish.

But perhaps most importantly, planning for their off-grid future required our family to grapple with the thorny issue of finances. As much as they longed for a life free from the constraints of money and materialism, our family understood that the practical realities of off-grid living demanded a certain degree of financial stability and foresight. So, they crunched the numbers, drafted budgets, and explored creative ways to fund their off-grid adventure, from saving diligently to seeking out grants and subsidies for renewable energy initiatives.

In the end, planning for their off-grid future was about more than just laying out a roadmap—it was about forging a vision, a dream of a life lived on their own terms, in harmony with the rhythms of nature and the whispers of the wild. It was about embracing the unknown with courage and conviction, and setting forth on a journey towards a future defined not by what they lacked, but by what they had—the boundless potential of their own resilience, ingenuity, and indomitable spirit.

Assessing Your Needs, Skills, and Resources

In the early stages of their off-grid journey, our family found themselves confronted with a daunting task: assessing their needs, skills, and resources in preparation for a life of self-sufficiency. It was a process fraught with uncertainty and self-discovery, yet it laid the foundation for their future success and resilience in the face of adversity.

Assessing your needs is the crucial first step towards off-grid living, requiring a careful examination of your daily habits, routines, and necessities. For our family, this meant taking stock of their energy consumption, water usage, and food requirements, as well as considering factors such as climate, terrain, and seasonal variations. By understanding their basic needs and requirements, they were able to tailor their off-grid infrastructure to suit their lifestyle and priorities, ensuring a smooth transition to their new way of life.

But assessing your needs is just the beginning. Equally important is an honest evaluation of your skills and abilities, as well as an exploration of the resources at your disposal. For our family, this meant identifying their strengths and weaknesses, from practical skills like carpentry and gardening to softer skills like problem-solving and communication. It also meant taking inventory of their physical and mental resources, from tools and equipment to financial assets and community connections.

Yet, perhaps the most valuable resource of all was their spirit of resilience and determination, their unwavering commitment to charting a new course towards self-sufficiency. Armed with a clear understanding of their needs, skills, and resources, our family set forth on their off-grid journey with confidence and purpose, ready to embrace whatever challenges and opportunities lay ahead.

As you embark on your own off-grid adventure, take the time to assess your needs, skills, and resources with care and intention. Consider not only the practical aspects of off-grid living but also the emotional and psychological dimensions. Reflect on your strengths and weaknesses, your passions and priorities, and use this knowledge to inform your decisions and actions as you carve out a life of self-sufficiency, resilience, and boundless possibility.

Setting Realistic Goals and Timelines

As our intrepid family embarked on their off-grid journey, they quickly realized the importance of setting realistic goals and timelines. In a world where instant gratification often reigns supreme, it can be tempting to dive headfirst into the wild unknown without a clear plan in mind. Yet, as our family discovered, such impulsive actions often lead to frustration and disappointment.

Setting realistic goals is not about limiting your aspirations or stifling your dreams—it's about laying a solid foundation for success, one that is built on careful planning, thoughtful consideration, and a healthy dose of pragmatism. It's about acknowledging your limitations, assessing your resources, and charting a course that is both ambitious and attainable.

For our family, setting realistic goals began with a candid assessment of their needs, skills, and resources. They took stock of their strengths and weaknesses, their financial situation, and their desired lifestyle, using this information to inform their goals and priorities. From there, they crafted a vision for their off-grid future—a roadmap that outlined their objectives, timelines, and milestones.

But setting realistic goals is only half the battle; equally important is establishing clear timelines for achieving them. In a world where time is often in short supply, it can be easy to procrastinate or become overwhelmed by the enormity of the task at hand. Yet, as our family discovered, breaking their goals down into manageable chunks and setting specific deadlines helped them stay focused and motivated, even in the face of adversity.

Of course, setting realistic goals and timelines is not without its challenges. It requires discipline, perseverance, and a willingness to adapt and evolve as circumstances change. Yet, for our family, the rewards far outweighed the risks. With each goal they achieved, they grew more confident in their abilities, more resilient in the face of adversity, and more determined to forge a life of self-sufficiency and abundance.

So, as you embark on your own off-grid journey, remember the importance of setting realistic goals and timelines. Take the time to assess your needs, skills, and resources, and craft a vision for your future that is both ambitious and attainable. And above all, stay true to your vision, stay focused on your goals, and never lose sight of the incredible potential that lies within you to thrive beyond the confines of civilization.

Choosing a Location and Considering Local Regulations

Selecting the perfect location for your off-grid homestead is a crucial step in your journey towards self-sufficiency. It's not just about finding a picturesque spot with breathtaking views; it's about carefully weighing various factors to ensure that your chosen location aligns with your goals, needs, and lifestyle preferences.

For our intrepid family, the process of choosing a location was a deeply personal and meticulously researched endeavour. They embarked on a quest to find a piece of land that offered the ideal balance of natural beauty, practicality, and regulatory feasibility. Here's how they navigated this crucial decision-making process:

1. **Assessing Your Needs and Preferences:** Before you start scouting for land, take some time to reflect on your needs, preferences, and priorities. Are you looking for a remote wilderness retreat, or do you prefer to be closer to civilization? What climate and terrain are most conducive to your desired lifestyle? Consider factors such as access to water sources, solar exposure, soil quality for gardening, and proximity to essential services like healthcare and groceries.

2. **Researching Potential Locations:** Armed with a clear understanding of your needs and preferences, begin researching potential locations that meet your criteria. Use online real estate listings, land databases, and local newspapers to identify promising properties in your desired area. Take note of key details such as acreage, topography, vegetation, and any existing infrastructure or amenities.

3. **Visiting Prospective Sites:** Once you've compiled a list of potential locations, it's time to hit the road and visit each site in person. This hands-on approach allows you to assess the land firsthand, soaking in its natural beauty and evaluating its suitability for your needs. Pay attention to factors such as accessibility, terrain features, natural resources, and the overall "feel" of the land.

4. **Consulting with Local Experts:** In addition to conducting your own research, don't hesitate to seek guidance from local experts who are familiar with the area. Reach out to real estate agents, land developers, and environmental consultants who can provide valuable insights into local land values, regulations, zoning restrictions, and environmental considerations. Their expertise can help you make informed decisions and avoid potential pitfalls.

5. **Considering Local Regulations:** One of the most critical aspects of choosing a location for your off-grid homestead is understanding and complying with local regulations and zoning ordinances. Different regions have varying laws regarding land use, building codes, environmental protection, and off-grid living practices. Before purchasing any property, thoroughly research the applicable regulations and seek clarification from local authorities if necessary.

6. **Balancing Freedom with Compliance**: While it's tempting to seek out remote, unregulated areas for off-grid living, it's essential to strike a balance between freedom and compliance with local regulations. Violating zoning laws or building codes can result in costly fines, legal battles, and even forced eviction from your property. Choose a location where you can live comfortably off-grid while remaining in compliance with applicable laws and regulations.

By following these steps and carefully considering all factors, our family successfully identified the perfect location for their off-grid homestead—a secluded parcel of land nestled in the embrace of nature, with abundant sunshine, fertile soil, and minimal regulatory restrictions. With their dream location secured, they were ready to embark on the next phase of their off-grid adventure, confident in their decision and excited for the challenges and opportunities that lay ahead.

Financial Considerations for Off-Grid Living

Embarking on the journey towards off-grid living is not merely a lifestyle choice; it's a significant financial investment that requires careful planning and consideration. As our intrepid family discovered, the path to self-sufficiency is paved with both challenges and opportunities, each one carrying its own financial implications.

At the outset of their journey, our family found themselves faced with the daunting task of assessing their financial resources and determining the feasibility of their off-grid dreams. From the cost of land acquisition to the expenses associated with infrastructure development, they quickly realized that off-grid living required a substantial upfront investment—one that would test their financial acumen and resourcefulness to the fullest.

But while the initial costs of off-grid living may seem daunting, our family soon discovered that the long-term savings and benefits far outweighed the upfront expenses. By harnessing renewable energy sources like solar power and wind energy, they were able to dramatically reduce their monthly utility bills, freeing up valuable resources that could be reinvested into their off-grid infrastructure.

Moreover, by growing their own food and implementing sustainable practices like composting and rainwater harvesting, our family was able to significantly reduce their monthly expenses while simultaneously improving their quality of life. No longer shackled by the endless cycle of consumerism and dependence on external resources, they found a newfound sense of freedom and empowerment in their off-grid lifestyle.

Of course, navigating the financial landscape of off-grid living requires careful planning and foresight. From budgeting for ongoing maintenance and repairs to setting aside funds for unexpected emergencies, it's essential to approach off-grid living with a realistic understanding of the financial commitments involved.

But for our family, the financial sacrifices were more than worth it in the end. As they watched their off-grid homestead flourish and thrive, they realized that the true value of self-sufficiency lies not just in the dollars and cents saved, but in the priceless sense of independence, resilience, and connection to the natural world that it brings.

So, as you consider embarking on your own off-grid journey, remember to approach it with both caution and optimism. While the financial considerations may seem daunting at first, the rewards of off-grid living—both tangible and intangible—are more than worth the investment. With careful planning, resourcefulness, and a willingness to embrace the challenges that lie ahead, you too can reap the countless benefits of off-grid living and embark on a journey towards a more sustainable, self-sufficient future.

Survival Projects for No Grid Living

Part 2

Building Your Off-Grid Infrastructure

CHAPTER 3
ENERGY GENERATION

In the heart of off-grid living, energy generation becomes the lifeblood of self-sufficiency. It's the power that fuels our dreams, the force that lights our path, and the cornerstone of our journey towards independence from the electrical grid. As our family ventured into the realm of energy generation, they quickly discovered that harnessing the power of the sun, wind, and biofuels was not just a practical necessity, but a profound act of stewardship towards the Earth and all its inhabitants.

Understanding Your Energy Needs

Before diving headlong into the world of energy generation, our family took a moment to pause and reflect on their unique energy needs. From lighting their home to powering essential appliances and tools, they carefully assessed their daily energy consumption, identifying areas where efficiency could be maximized and waste minimized. By taking stock of their energy demands, they were better equipped to tailor their off-grid energy system to meet their specific requirements, ensuring that every watt generated was put to good use.

Solar Power: Harnessing the Power of the Sun

In the vast expanse of off-grid living, few resources are as abundant and reliable as the sun. For our intrepid family, harnessing the power of solar energy was not just a practical necessity but a profound testament to the ingenuity and resilience of the human spirit. As they embarked on their journey towards self-sufficiency, they quickly realized that the sun, with its boundless energy and unwavering presence, would be their most faithful ally in their quest for off-grid resilience.

Imagine their delight as they set up their first solar panel array, watching in awe as the sun's rays danced across the surface of the photovoltaic cells, transforming sunlight into electricity with astonishing efficiency. Here, amidst the quiet beauty of the wilderness, they found a source of power that was not just clean and sustainable but utterly limitless—a gift from the cosmos that would sustain them through sunlit days and starry nights alike.

But harnessing the power of the sun was no simple task. It required careful planning, meticulous design, and a deep understanding of the principles of solar energy. From calculating their energy needs to sizing their solar system accordingly, our family embarked on a journey of discovery, learning to navigate the intricate world of solar power with skill and precision.

Yet, for all its complexity, solar power offered a wealth of benefits that far outweighed the challenges. In a world grappling with environmental crisis and dwindling fossil fuel reserves, solar energy emerged as a beacon of hope—a clean, renewable resource that promised to power our homes, businesses, and communities for generations to come.

One of the greatest advantages of solar power was its versatility. From small-scale off-grid systems to large-scale grid-tied installations, solar energy could be harnessed in a myriad of ways to meet a wide range of energy needs. Whether it was powering lights and appliances in a remote cabin or providing electricity to an entire village, the sun's energy offered a flexible and reliable solution to the pressing energy challenges of our time.

But perhaps the most profound benefit of solar power was its environmental impact. Unlike fossil fuels, which release harmful greenhouse gases and contribute to climate change, solar energy is clean, green, and infinitely sustainable. By harnessing the power of the sun, our family not only reduced their carbon footprint but also took a bold step towards building a more sustainable future for themselves and for generations to come.

In the end, solar power was more than just a source of electricity—it was a symbol of hope, a testament to the power of human ingenuity, and a beacon of light in a world shrouded in darkness. As our family basked in the warm glow of the sun's rays, they knew that they had embarked on a journey that would not just sustain them but inspire others to follow in their footsteps, harnessing the power of the sun to create a brighter, more sustainable world for all.

Wind Power: Harnessing the Power of the Wind

In the vast expanse of the wilderness, where the air is crisp and the horizon stretches endlessly, lies an untapped source of boundless energy—the wind. For our intrepid family venturing into the realm of off-grid living, harnessing the power of the wind was not just a practical necessity but a profound expression of their commitment to sustainability and self-sufficiency.

Imagine the scene: atop a windswept hill, overlooking a panorama of rolling hills and majestic forests, our family stands in awe of the raw power swirling around them. With each gust, they feel the gentle caress of the wind against their skin, a reminder of nature's awesome might and infinite potential.

But harnessing the power of the wind is no simple feat. It requires ingenuity, creativity, and a deep understanding of the forces at play. For our family, it began with a humble wind turbine—a towering sentinel standing tall against the backdrop of the sky, its blades slicing through the air with graceful precision.

As they set to work assembling their wind turbine, our family marvelled at the elegance of its design—the way its sleek blades captured the energy of the wind and transformed it into electricity, ready to power their off-grid homestead with clean, renewable energy. With each turn of the blades, they felt a sense of pride and accomplishment, knowing that they were harnessing the power of nature to fuel their journey towards self-sufficiency.

But wind power is more than just a practical solution to off-grid energy needs—it's a symbol of our interconnectedness with the natural world, a reminder that we are but humble stewards of the Earth's resources. As our family gazed out across the landscape, they felt a deep sense of gratitude for the wind that breathed life into their off-grid oasis, powering their dreams and lighting their way towards a brighter, more sustainable future.

Of course, harnessing the power of the wind comes with its own set of challenges. From unpredictable weather patterns to the occasional technical glitch, our family quickly learned that living off-grid required adaptability and resilience in the face of adversity. Yet, with each challenge they encountered, they grew stronger and more determined, their commitment to sustainability unwavering in the face of uncertainty.

In the end, the rewards of harnessing the power of the wind far outweighed the challenges. From the thrill of watching their turbine come to life in the breeze to the satisfaction of knowing that they were reducing their carbon footprint and protecting the planet for future generations, wind power became more than just a source of energy—it became a symbol of hope, a beacon of light in a world grappling with the consequences of climate change.

So, as you embark on your own journey towards self-sufficiency, remember the power of the wind—the gentle whisper of nature's song, calling us to embrace a future powered by clean, renewable energy. With determination, ingenuity, and a deep respect for the Earth, you too can harness the power of the wind and pave the way towards a brighter, more sustainable tomorrow.

Biofuel Power: Harnessing the Power of Biomass

In the realm of off-grid living, harnessing the power of biofuels offers a renewable and sustainable alternative to traditional energy sources. For our intrepid family, the journey into biofuel power was a testament to their ingenuity and commitment to living in harmony with the Earth. As they delved into the world of biomass, they discovered a wealth of potential waiting to be unlocked—a hidden reservoir of energy waiting to be harnessed for the benefit of all.

Biomass, in its simplest form, refers to organic matter derived from living or recently living organisms. This can include everything from agricultural residues and forestry byproducts to dedicated energy crops grown specifically for fuel production. By converting this organic matter into biofuels, our family found a renewable and environmentally friendly source of energy that could be used to power everything from their homes to their vehicles.

One of the most common forms of biofuel is biodiesel, a renewable fuel made from vegetable oils or animal fats. For our family, biodiesel offered a cleaner and more sustainable alternative to traditional diesel fuel, allowing them to power their vehicles and machinery without relying on fossil fuels. By sourcing their feedstock from locally available resources, they were able to reduce their carbon footprint while supporting local agriculture and industry.

But biodiesel was just the beginning. As they delved deeper into the world of biofuel production, our family discovered a myriad of other options waiting to be explored. From biogas produced through anaerobic digestion of organic waste to ethanol derived from fermenting sugars and starches, the possibilities were endless. Each biofuel offered its own unique set of advantages and challenges, but all shared a common goal—to provide a renewable and sustainable source of energy for off-grid living.

Of course, harnessing the power of biomass was not without its challenges. From sourcing feedstock to processing and refining biofuels, our family faced numerous obstacles along the way. Yet, with determination and ingenuity, they were able to overcome these challenges and unlock the full potential of biofuel power.

One of the key advantages of biofuel power is its versatility. Unlike traditional energy sources, which are often centralized and dependent on complex infrastructure, biofuels can be produced on a small scale using relatively simple equipment. This makes them particularly well-suited for off-grid living, where access to traditional energy sources may be limited or unreliable.

But perhaps the greatest advantage of biofuel power is its sustainability. Unlike fossil fuels, which are finite and contribute to climate change, biofuels are renewable and carbon-neutral, meaning they release no more carbon dioxide into the atmosphere than was originally absorbed by the plants from which they are derived. This makes them a vital tool in the fight against climate change and a key component of a sustainable energy future.

In the end, harnessing the power of biomass was more than just a practical choice for our family—it was a commitment to living in harmony with the Earth, to reducing their environmental impact, and to forging a brighter, more sustainable future for generations to come. Through their journey into biofuel power, they discovered a renewable and sustainable source of energy that empowered them to live off the grid with confidence and pride.

Micro Hydropower: Harnessing the Power of Flowing Water

In the heart of the wilderness, where rivers carve their winding paths through the rugged terrain, lies a hidden source of untapped energy waiting to be harnessed. For our intrepid family, micro hydropower became more than just a renewable energy source—it became a symbol of their commitment to sustainable living, a testament to their ingenuity, and a beacon of hope in a world grappling with environmental crisis.

Imagine their excitement as they stumbled upon a babbling brook, its crystal-clear waters cascading over moss-covered rocks, a symphony of sound and motion that spoke of untold power waiting to be unleashed. Here, amidst the tranquil beauty of nature, they saw not just a stream, but a potential source of clean, renewable energy—a lifeline that could power their off-grid oasis and fuel their dreams of self-sufficiency.

But harnessing the power of flowing water was no small feat. It required careful planning, meticulous engineering, and a deep understanding of the principles of hydrodynamics. From designing intake structures to building penstocks and turbines, every step of the process demanded precision and expertise, lest their ambitious project be swept away by the relentless force of the river.

Yet, as they delved deeper into the world of micro hydropower, our family discovered a wealth of resources and support waiting to guide them on their journey. From technical manuals and online forums to local experts and renewable energy cooperatives, they found a community of like-minded individuals eager to share their knowledge and expertise, helping them to navigate the complexities of hydroelectricity with confidence and grace.

With each passing day, their micro hydropower system took shape, transforming the natural flow of the river into a steady stream of electricity that powered their off-grid homestead and fuelled their dreams of a more sustainable future. From lighting their home to running their appliances, their micro hydropower system became the beating heart of their off-grid oasis, a constant reminder of the power of nature and the boundless potential of renewable energy.

But perhaps the greatest reward of all was the knowledge that their micro hydropower system was more than just a source of energy—it was a symbol of their commitment to living in harmony with the Earth, a testament to their belief in a future powered by clean, renewable energy.

In harnessing the power of flowing water, our family had not just built a micro hydropower system—they had built a brighter, more sustainable future for themselves and for generations to come.

CHAPTER 4
WATER COLLECTION AND PURIFICATION

In the vast expanse of off-grid living, water is life—a precious resource that sustains us, nourishes us, and connects us to the rhythm of the natural world. As our intrepid family embarked on their journey towards self-sufficiency, they quickly realized the critical importance of water collection and purification, mastering the art of harnessing nature's bounty while ensuring its purity for consumption and use.

Water Needs for Drinking, Sanitation, and Irrigation

Water—the elixir of life, the essential ingredient for survival in the wild embrace of off-grid living. As our family settled into their new home amidst the rugged beauty of the wilderness, they quickly learned to cherish every precious drop, treating it not just as a resource, but as a sacred gift from the Earth.

In the realm of off-grid living, water serves a multitude of purposes, each one essential for sustaining life and fostering resilience. From quenching our thirst to nourishing our crops, from maintaining sanitation to supporting hygiene, water is the lifeblood of our existence, a fundamental cornerstone of off-grid survival.

Drinking Water:

In the early days of their off-grid journey, our family's first priority was ensuring a clean, reliable source of drinking water. They embarked on a quest to harness the bounty of nature, seeking out pristine springs and mountain streams, where crystal-clear water flowed freely from the earth. With careful planning and meticulous filtration, they transformed these natural springs into a reliable source of pure, refreshing water, ensuring that their family stayed hydrated and healthy even in the most remote corners of the wilderness.

But as they soon discovered, access to clean drinking water wasn't always a given in the wilds of off-grid living. In times of drought or extreme weather, their once-abundant water sources could dwindle to a trickle, leaving them vulnerable to dehydration and illness. It was during these challenging times that they learned the true value of conservation and resourcefulness, treating every drop of water as a precious commodity to be cherished and preserved.

Sanitation:

Beyond quenching their thirst, water played a vital role in maintaining sanitation and hygiene in their off-grid homestead. With no municipal sewage system to rely on, our family had to devise innovative solutions for managing waste and minimizing environmental impact. They embraced the principles of eco-friendly living, implementing composting toilets and greywater recycling systems to ensure that every drop of water was used wisely and responsibly.

But sanitation wasn't just about managing waste—it was also about safeguarding their health and well-being in the face of potential hazards. In a world where clean water was not always guaranteed, our family took proactive measures to purify and filter their water supply, investing in high-quality filtration systems and UV sterilizers to ensure that every sip was safe and clean.

Irrigation:

As our family's off-grid homestead flourished and expanded, so too did their need for water to support their burgeoning garden and orchard. With careful planning and strategic design, they implemented a sustainable irrigation system that maximized efficiency and minimized waste, harnessing the power of gravity and rainwater harvesting to nourish their crops and sustain their livelihood.

But irrigation wasn't just about watering their plants—it was also about cultivating a deeper connection to the land and fostering a sense of stewardship for the earth. As they tended to their garden with tender care and loving attention, our family found solace and fulfillment in the simple act of nurturing new life, knowing that each drop of water was a gift from the earth, a symbol of abundance and renewal in the wild embrace of off-grid living.

In the realm of off-grid survival, water is more than just a resource—it's a lifeline, a beacon of hope in a world fraught with uncertainty. So, as you embark on your own off-grid journey, remember to cherish every drop, to conserve and protect this precious gift from the earth, for in its embrace lies the key to survival and resilience in the wilds of off-grid living.

Rainwater Harvesting

As the clouds gathered overhead and the first droplets of rain began to fall, our adventurers sprang into action, harnessing the power of nature to meet their water needs. They quickly learned the art of rainwater harvesting, employing a variety of ingenious techniques to capture and store this precious resource for future use.

From rooftop catchment systems to simple rain barrels nestled beneath the eaves, our adventurers utilized every available surface to maximize their water collection efforts. With each passing storm, their reservoirs swelled with the bounty of the heavens, providing a reliable source of clean, fresh water to sustain them through the dry spells ahead.

But rainwater harvesting was just the beginning; our adventurers soon realized that proper filtration was essential to ensure the safety and purity of their water supply. They invested in high-quality filtration systems, carefully designed to remove impurities and contaminants, ensuring that every drop that passed their lips was as pure as the mountain spring from which it flowed.

Greywater Systems

In the intricate web of off-grid living, every drop of water is a precious resource—a lifeline that sustains not only our bodies but also the delicate balance of the ecosystem around us. And when it comes to maximizing water efficiency, few systems are as ingenious as greywater systems.

Imagine the gentle trickle of water from your kitchen sink or the soothing flow from your shower—a seemingly mundane sight, yet brimming with untapped potential. Greywater, often overlooked and underutilized, is the key to unlocking a world of sustainable water management, where every drop is cherished, recycled, and repurposed to nourish the earth and sustain life.

But what exactly is greywater, and how does it differ from its counterpart, blackwater? Greywater refers to wastewater generated from sources such as sinks, showers, and washing machines—water that is relatively clean and free from harmful contaminants. Unlike blackwater, which contains sewage and requires extensive treatment, greywater can be safely reused for a variety of non-potable purposes, from irrigation to toilet flushing.

For our intrepid family, greywater systems became a cornerstone of their off-grid lifestyle—a simple yet ingenious solution to the perennial challenge of water conservation. With a few basic components and a bit of ingenuity, they transformed their humble abode into a model of water efficiency, recycling greywater to nourish their gardens, replenish their aquifers, and reduce their reliance on precious freshwater sources.

But the benefits of greywater systems extend far beyond water conservation alone. By diverting greywater from traditional sewage systems, our family reduced the burden on municipal infrastructure, alleviated strain on local ecosystems, and mitigated the risk of pollution in nearby waterways. In doing so, they not only reclaimed control over their water supply but also forged a deeper connection with the land that sustained them—a connection rooted in reciprocity, respect, and reverence for the natural world.

Of course, implementing a greywater system is not without its challenges. From regulatory hurdles to technical complexities, there are many factors to consider when designing and installing a greywater system. But with proper planning, guidance, and a willingness to innovate, our family discovered that the rewards far outweighed the risks—a life where water flows freely, sustainably, and abundantly, nourishing not only their bodies but also their souls.

So, as you embark on your own off-grid journey, consider the humble greywater system—a beacon of hope in a world grappling with water scarcity and environmental degradation. With its simple elegance and profound impact, it offers a glimpse into a future where water is cherished, recycled, and revered—a future where every drop truly matters.

In the quest for self-sufficiency, securing a reliable source of clean water is paramount. Our intrepid family understood this fundamental truth as they embarked on their off-grid journey, recognizing the vital importance of harnessing nature's gifts while ensuring the safety and purity of their water supply. Amidst the lush beauty of their wilderness sanctuary, they explored a variety of natural water purification methods, each one offering a time-honoured solution to the age-old challenge of water sanitation.

Natural Water Purification Methods

As our family ventured deeper into the heart of off-grid living, they discovered a wealth of natural water purification methods, each one rooted in centuries-old wisdom and grounded in the immutable laws of nature. From ancient filtration techniques to time-tested purification rituals, these methods offered a sustainable and eco-friendly alternative to conventional water treatment systems, harnessing the power of Mother Earth to transform even the murkiest of waters into a source of pristine refreshment.

Filtration Techniques:

At the forefront of natural water purification lies the art of filtration, a process that relies on the physical properties of various materials to remove impurities from water. Our family experimented with a range of filtration mediums, from coarse sand to fine charcoal, each one meticulously layered to create a makeshift filtration system that rivalled even the most advanced water treatment plants. As water passed through these layers, contaminants were trapped and neutralized, leaving behind a crystal-clear liquid that sparkled with the purity of mountain springs.

Distillation and Boiling:

In their pursuit of pure water, our family turned to the age-old techniques of distillation and boiling, methods that have stood the test of time as reliable means of water purification. By heating water to its boiling point and capturing the steam, they were able to separate out impurities and pathogens, leaving behind a potable liquid that was safe for drinking and cooking. It was a simple yet effective solution, one that required little more than a fire and a sturdy pot, yet yielded results that were nothing short of miraculous.

Natural Filtration by Plants:

In their exploration of natural water purification methods, our family stumbled upon a hidden gem—a lush oasis teeming with aquatic plants known for their remarkable ability to filter and purify water. From the delicate fronds of water hyacinths to the sturdy stems of cattails, these plants acted as nature's own filtration system, absorbing toxins and pollutants from the water and releasing clean, oxygen-rich liquid in their wake. Inspired by this natural marvel, our family set about cultivating their own mini wetland, harnessing the power of plants to transform their pond into a haven of purity and abundance.

As our family immersed themselves in the world of natural water purification, they were struck by the profound simplicity and elegance of these age-old techniques. From filtration to distillation, nature offered a bounty of solutions to the timeless challenge of water sanitation, reminding us that in the dance of life, the answers we seek are often found in the quiet whispers of the natural world. So, as you embark on your own off-grid journey, take heart in the knowledge that the wisdom of the ages is at your fingertips, waiting to guide you on your quest for clean, pure water in even the most remote corners of the earth.

CHAPTER 5
FOOD PRODUCTION AND PRESERVATION

In the heart of self-sufficiency lies the ability to cultivate and harvest nourishment from the land. Our family's journey towards food production and preservation was not just about sustenance—it was a profound connection to the earth, a celebration of the cycles of life, and a testament to the power of resilience in the face of adversity.

As they surveyed their new homestead, our family knew that food production would be the cornerstone of their off-grid life. With determination and ingenuity, they set out to create a thriving oasis of abundance, where the bounty of the earth would sustain them through the changing seasons and the trials of off-grid living.

Planning Your Off-Grid Garden:

In the heart of every off-grid homestead lies a thriving garden—a verdant oasis teeming with life, sustenance, and boundless potential. As our intrepid family set out to cultivate their own off-grid garden, they quickly discovered that planning was key to success. From selecting the right crops to designing efficient irrigation systems, every decision played a crucial role in shaping their garden into a flourishing haven of self-sufficiency.

Choosing the Right Vegetables and Fruits

The first step in planning your off-grid garden is selecting the right mix of vegetables and fruits to suit your climate, soil, and dietary preferences. Our family began by conducting thorough research into native plant species and heirloom varieties, seeking out crops that were well-suited to their region's growing conditions and nutritional needs.

They carefully considered factors such as sunlight exposure, soil quality, and water availability, choosing crops that thrived in their particular microclimate. From hearty root vegetables like potatoes and carrots to leafy greens like kale and spinach, their garden was a diverse tapestry of colours, flavours, and textures—a testament to the bounty of nature's harvest.

Seed Selection and Starting Seedlings

With their crop selection finalized, our family turned their attention to sourcing high-quality seeds and starting healthy seedlings. They scoured seed catalogues and online retailers for heirloom varieties renowned for their flavour, resilience, and adaptability, prioritizing seeds from reputable suppliers committed to sustainable growing practices.

Armed with their carefully chosen seeds, they set up a cozy corner of their cabin for seed starting, equipped with grow lights, seed trays, and organic potting soil. With tender care and attention, they nurtured their seedlings from tiny sprouts to robust plants, ensuring that each one was strong and healthy before transplanting them into the garden beds.

Organic Gardening Practices

Central to the success of their off-grid garden was a commitment to organic gardening practices. Our family eschewed synthetic pesticides and fertilizers in favour of natural alternatives like compost, mulch, and companion planting. They embraced permaculture principles, creating a harmonious ecosystem where plants, animals, and beneficial insects coexisted in perfect balance.

They employed techniques like crop rotation and intercropping to maximize space and minimize pests and diseases, cultivating a diverse array of plants that complemented and supported each other's growth. Their garden was a vibrant tapestry of biodiversity, teeming with pollinators, beneficial insects, and soil microbes—a living testament to the power of organic gardening to nourish both body and soul.

Harvesting and Preservation

As the seasons turned and their garden burst into full bloom, our family revelled in the joy of harvest time—a time of abundance, celebration, and gratitude. They carefully tended to their crops, harvesting them at peak ripeness to ensure maximum flavour and nutrition. They savoured the simple pleasure of plucking ripe tomatoes from the vine, crisp cucumbers

from the trellis, and sweet berries from the bramble—a feast for the senses and a balm for the soul.

But the bounty of the garden didn't end with the harvest. Our family embraced the art of food preservation, employing techniques like canning, drying, and fermentation to extend the shelf life of their harvest and enjoy its bounty year-round. They filled their pantry shelves with jars of homemade pickles, bags of dried herbs, and crocks of sauerkraut, ensuring that not a single precious morsel went to waste.

In their off-grid garden, our family found not just sustenance, but solace—a sanctuary where they could reconnect with the rhythms of nature, nurture their bodies and souls, and cultivate a deep sense of gratitude for the gifts of the Earth. And as they tended to their garden with love and care, they discovered that the truest measure of wealth lies not in material possessions, but in the rich abundance of the natural world that surrounds us.

Raising Small Livestock

As our family settled into their off-grid homestead, they recognized the importance of diversifying their sources of sustenance. While their garden provided an abundance of fresh fruits and vegetables, they understood that true self-sufficiency required a more comprehensive approach—one that included raising small livestock.

Picture the scene: A cozy corner of the homestead transformed into a bustling mini-farm, alive with the sounds of clucking chickens, contented goats, and the occasional oink of a piglet. For our family, raising small livestock wasn't just a practical necessity; it was a labour of love, a daily reminder of their deep connection to the land and its inhabitants.

At the heart of their mini-farm were the chickens—feathered friends who provided a steady supply of fresh eggs, rich in protein and nutrients. Our family delighted in the daily ritual of collecting eggs from the coop, marvelling at the vibrant hues of their shells and savouring the rich, golden yolks that awaited them inside.

But chickens were just the beginning. Our family also welcomed a small herd of goats onto their homestead, recognizing their dual role as providers of milk and companionship. With gentle care and affection, they tended to their goats, milking them each morning and evening and using the rich, creamy milk to create an array of dairy delights, from cheese and yogurt to butter and ice cream.

And then there were the pigs—curious, intelligent creatures who roamed the fields with boundless energy and enthusiasm. Our family delighted in watching them root and forage, their snouts digging up treasures from the earth below. And when the time came, they bid a bittersweet farewell to their porcine friends, grateful for the nourishment they provided and the lessons they taught about the cycle of life and death.

Raising small livestock was more than just a practical endeavour for our family; it was a profound expression of their commitment to self-sufficiency and sustainability. With each egg, each glass of milk, each slice of bacon, they were reminded of the interconnectedness of all living things and the importance of treating the earth and its creatures with respect and gratitude.

So, as you consider embarking on your own journey of raising small livestock, remember to approach it with an open heart and a spirit of reverence. Whether you're tending to chickens, goats, pigs, or any other small animals, treat them with kindness and care, and they will reward you with an abundance of sustenance, companionship, and joy.

Food Preservation Techniques:

In the lush embrace of their off-grid homestead, our intrepid family discovered the true joy of self-sufficiency—one that extended far beyond the mere act of growing their own food. For them, it was not just about cultivating a bountiful garden or raising livestock; it was about preserving the fruits of their labour, ensuring that the abundance of summer could sustain them through the long winter months.

As they surveyed their harvest with pride and gratitude, our family quickly realized the importance of mastering the art of food preservation. With careful planning and foresight, they set out to explore a myriad of techniques, each one offering a unique way to capture the fleeting essence of summer and preserve it for the months ahead.

Canning: Armed with mason jars and a trusty pressure canner, our family embarked on their journey into the world of canning—a time-honoured tradition that transforms fresh produce into pantry staples that can last for years. From vibrant jams and jellies to savory pickles and relishes, they discovered the endless possibilities that canning offered, each jar a testament to their hard work and dedication.

Drying: With the gentle warmth of the sun as their ally, our family embraced the ancient art of drying, transforming surplus fruits and vegetables into nutrient-dense snacks that would sustain them through the lean times. From sun-dried tomatoes and apricots to crispy kale chips and jerky, they marvelled at the simplicity and versatility of this age-old technique, each bite a taste of summer's bounty.

Pickling: In the heart of their cozy kitchen, our family revelled in the tangy delights of pickling, as they transformed crisp cucumbers, crunchy carrots, and plump peppers into a rainbow of vibrant delights. With a medley of spices and herbs at their disposal, they experimented with flavours and combinations, each jar a testament to their creativity and ingenuity.

Fermentation: As they delved deeper into the world of food preservation, our family discovered the magic of fermentation—a process that not only preserves food but also enhances its flavour and nutritional value. From tangy sauerkraut and kimchi to fizzy kombucha and sourdough bread, they marvelled at the transformative power of beneficial bacteria, each batch a celebration of life's natural rhythms.

Root Cellaring: And finally, as the harvest season drew to a close, our family prepared to store their bounty for the long winter ahead. With a sturdy root cellar as their sanctuary, they carefully packed away their treasures, from crisp apples and sweet potatoes to hearty squash and onions. In the cool darkness of their underground sanctuary, their harvest would slumber peacefully, waiting to nourish them through the cold months ahead.

As they surveyed their well-stocked pantry and cellar with pride and satisfaction, our family knew that they were prepared for whatever challenges lay ahead. With a wealth of preserved food at their fingertips, they could face the future with confidence, knowing that they had mastered the art of food preservation and embraced the timeless wisdom of self-sufficiency.

Part 3

Shelter and Waste Management

CHAPTER 6
MAKING YOUR HOME MORE ENERGY-EFFICIENT

As our family settled into their off-grid homestead, they quickly realized that making their home more energy-efficient was not just a luxury—it was a necessity. With limited resources at their disposal, they knew that every watt of electricity saved was a step closer to true self-sufficiency.

Their journey towards energy efficiency began with a thorough assessment of their home's energy consumption. Armed with a keen eye and a notepad, they meticulously inspected every nook and cranny, identifying energy leaks and inefficiencies with surgical precision. From drafty windows to poorly insulated walls, no detail escaped their scrutiny as they sought to maximize their home's energy efficiency.

With their list of improvements in hand, our family set to work implementing a series of cost-effective upgrades designed to slash their energy consumption and reduce their environmental footprint. They started by beefing up their home's insulation, sealing cracks and crevices with weatherstripping and caulking to prevent heat loss and keep their living spaces cozy and warm.

Next, they turned their attention to their heating and cooling systems, investing in energy-efficient appliances and implementing passive heating and cooling techniques to minimize their reliance on fossil fuels. They installed ceiling fans to circulate air more efficiently, strategically placed shade trees to block out the sun's harsh rays in the summer, and installed programmable thermostats to optimize their heating and cooling schedules based on their daily routines.

But perhaps the most transformative upgrade of all was their decision to embrace renewable energy sources, harnessing the power of the sun and wind to meet their home's energy needs. They installed a robust solar panel array on their roof, basking in the warm glow of clean, renewable energy as it flowed into their home, powering their lights, appliances, and electronics with the boundless energy of the sun.

Outside, a sleek wind turbine spun gracefully in the breeze, harnessing the power of the wind to generate electricity and supplement their solar power system on cloudy days and calm nights. Together, these renewable energy sources formed the backbone of their off-grid energy infrastructure, providing them with a reliable and sustainable source of power that freed them from the constraints of the grid.

But their quest for energy efficiency didn't end there. Our family continued to explore innovative ways to reduce their energy consumption and live more sustainably, from installing energy-efficient LED light bulbs to embracing low-tech solutions like passive solar heating and daylighting. With each new improvement, they drew closer to their goal of true self-sufficiency, forging a path towards a brighter, more sustainable future for themselves and future generations to come.

So, as you embark on your own journey towards energy efficiency, take inspiration from our family's story and embrace the power of innovation, resourcefulness, and determination. With a little creativity and a lot of perseverance, you too can transform your home into an energy-efficient oasis, reducing your environmental footprint and embracing a life of true self-sufficiency in the wild embrace of off-grid living.

Identifying Energy Leaks in Your Home

As our intrepid family settled into their off-grid lifestyle, they quickly realized that achieving energy efficiency was not just a matter of generating renewable power—it was also about minimizing waste and optimizing their existing resources. And so, they embarked on a mission to identify and eliminate energy leaks in their humble abode, transforming their cozy cabin into a model of efficiency and sustainability.

But where to begin? For our family, the journey started with a simple yet powerful realization: energy leaks can hide in plain sight, lurking in the shadows of our daily routines and unnoticed corners of our homes. From drafty windows to poorly insulated walls, every nook and cranny held the potential for wasted energy and lost savings.

Armed with determination and a keen eye for detail, our family set out to uncover these hidden energy leaks, starting with a thorough inspection of their home's exterior. They traced the outline of every window and door, feeling for telltale drafts and gaps that betrayed the presence of unwanted airflow. They scrutinized the walls for signs of moisture and mould, indicators of compromised insulation and structural integrity.

But the real breakthrough came when they turned their attention inward, peering beneath the surface of their home to uncover the secrets of its inner workings. They inspected the ductwork for signs of leakage, sealing any cracks and gaps with precision and care. They checked the insulation in the attic and crawl spaces, ensuring that every square inch was properly fortified against the elements.

Yet, perhaps the most eye-opening discovery came when they examined their everyday appliances and electrical fixtures. They replaced outdated incandescent light bulbs with energy-efficient LEDs, reducing their lighting costs while illuminating their home with a warm, inviting glow. They installed programmable thermostats and smart power strips, allowing them to effortlessly control their energy usage and minimize standby power consumption.

But the journey didn't end there. With each energy leak they identified and sealed, our family found themselves inspired to delve deeper, to explore new ways of maximizing their home's efficiency and minimizing their environmental impact. They experimented with passive heating and cooling techniques, harnessing the power of sunlight and natural ventilation to maintain a comfortable indoor climate year-round.

And as they stood back and surveyed their handiwork, our family couldn't help but feel a sense of pride and accomplishment. In the span of just a few short weeks, they had transformed their humble cabin into a beacon of efficiency and sustainability, a testament to the power of ingenuity and determination in the face of adversity.

So, as you embark on your own journey towards energy efficiency, take heart in knowing that every small step you take brings you one step closer to a brighter, more sustainable future. Whether you're sealing drafty windows or upgrading your appliances, remember that every action counts, and that together, we have the power to make a difference in the world around us.

Improving Insulation and Air Sealing

As our intrepid family settled into their off-grid abode, they quickly realized the importance of making their home as energy-efficient as possible. In the heart of the wilderness, where every watt of electricity and drop of water was precious, maximizing the efficiency of their living space became not just a luxury but a necessity.

One of the first steps they took towards achieving this goal was improving insulation and air sealing. In a world where winter temperatures could plummet to bone-chilling lows and summer heatwaves could leave them sweltering in the sun, ensuring that their home was properly insulated and airtight was paramount to their comfort and survival.

With meticulous attention to detail, they set about sealing every crack and crevice, every nook and cranny where precious heat could escape or unwanted drafts could creep in. They caulked windows and doors, installed weather-stripping around frames, and sealed gaps in the walls with expanding foam insulation. They even went so far as to insulate their floors and ceilings, creating a veritable cocoon of warmth and comfort in the heart of the wilderness.

But their efforts didn't stop there. Recognizing that proper ventilation was just as important as insulation, they installed energy-efficient windows and skylights that allowed natural light to flood their living space while minimizing heat loss. They also invested in a high-quality ventilation system that provided fresh air without compromising their home's energy efficiency, ensuring that they could breathe easy even in the depths of winter.

The results of their efforts were nothing short of remarkable. Gone were the days of shivering beneath layers of blankets or sweltering in the oppressive heat of summer. Instead, they found themselves basking in the warmth of a cozy fire during the winter months and enjoying the cool breeze of a summer evening without fear of energy waste or discomfort.

But perhaps the most profound impact of their energy-efficient home was felt not in their physical comfort but in their peace of mind. Knowing that they were minimizing their environmental footprint, conserving precious resources, and living in harmony with the natural world filled them with a sense of purpose and fulfillment that no number of modern conveniences could ever replicate.

So, as you embark on your own journey towards energy efficiency, remember the lessons learned by our intrepid family. By improving insulation and air sealing, you're not just creating a more comfortable living space—you're forging a deeper connection to the world around you, one that nourishes your soul as much as it warms your body.

In the ever-evolving landscape of off-grid living, mastering passive heating and cooling techniques is akin to unlocking the secrets of nature's thermostat. It's about harnessing the power of sunlight, airflow, and thermal mass to create a comfortable and sustainable living environment, without the need for costly mechanical systems or excessive energy consumption.

Picture this: Our intrepid family, nestled snugly in their off-grid cabin, basks in the warmth of the morning sun filtering through the south-facing windows. As the day progresses and temperatures rise, they draw upon the principles of passive cooling to keep their home comfortable and inviting, even in the sweltering heat of summer.

Passive heating begins with thoughtful design and orientation, positioning the home to maximize solar gain during the winter months while minimizing exposure to the harsh summer sun. South-facing windows capture the low-angle winter sunlight, warming the interior through radiant heat, while strategically placed overhangs and deciduous trees provide shade during the hotter months, reducing the need for artificial cooling.

But passive heating isn't just about capturing sunlight—it's also about retaining that heat through efficient insulation and thermal mass. Thick walls, insulated floors, and thermal curtains help to trap warmth inside the home, while materials like adobe, stone, or rammed earth absorb and store heat during the day, releasing it slowly at night to maintain a comfortable temperature without relying on external heating sources.

Similarly, passive cooling relies on natural ventilation and thermal mass to regulate indoor temperatures without the need for air conditioning. Cross-ventilation, achieved through strategically placed windows and vents, allows cool breezes to flow through the home, carrying away excess heat and humidity. Meanwhile, thermal mass, such as earth berms or concrete floors, absorbs and dissipates heat, keeping the interior cool and comfortable even on the hottest days.

For our family, mastering passive heating and cooling techniques was a game-changer, transforming their off-grid cabin into a haven of comfort and sustainability. From cozy winter nights by the fire to breezy summer afternoons on the porch, they found joy and contentment in the simple pleasures of off-grid living, all while minimizing their impact on the environment and reducing their reliance on fossil fuels.

So, as you embark on your own off-grid journey, remember that the power to heat and cool your home lies not in expensive gadgets or elaborate systems, but in the timeless wisdom of passive design. By harnessing the natural forces of sunlight, airflow, and thermal mass, you too can create a space that is both comfortable and sustainable, allowing you to thrive in harmony with the rhythms of nature, now and for generations to come.

Energy-Efficient Appliances

In the quest for off-grid living, energy efficiency is paramount. Every watt saved is a step closer to self-sufficiency, a testament to our commitment to preserving the delicate balance of our planet's resources. And at the heart of this endeavour lies the humble yet mighty energy-efficient appliance—a silent champion in the battle against waste and excess.

Imagine stepping into a home where every appliance hums with efficiency, where every watt is precious and every resource cherished. In our family's off-grid haven, energy-efficient appliances weren't just a luxury—they were a lifeline, a lifeline that ensured their comfort and sustainability in a world driven by consumption and waste.

At the heart of their energy-efficient arsenal lay the refrigerator—a marvel of modern engineering designed to keep their food fresh and their energy bills low. Gone were the days of bulky, power-hungry fridges; in their place stood sleek, energy-efficient models, meticulously engineered to minimize energy consumption without sacrificing performance. With advanced insulation, variable-speed compressors, and smart temperature controls, these refrigerators kept their food cool and their energy usage in check, ensuring that every watt was put to good use.

But the refrigerator was just the beginning. In their off-grid oasis, every appliance—from the humble light bulb to the mighty washing machine—was chosen with energy efficiency in mind. LED light bulbs illuminated their home with a warm, inviting glow while sipping power sparingly, ensuring that every room was bathed in light without draining their precious energy reserves. Their washing machine was a marvel of efficiency, using advanced technology to clean their clothes with minimal water and energy, leaving them fresh and clean without taxing their off-grid systems.

Yet, perhaps the most remarkable aspect of their energy-efficient appliances was their ability to adapt and evolve with the changing seasons. In the long, sun-drenched days of summer, their air conditioner hummed softly in the background, keeping their home cool and comfortable without driving up their energy bills. And when winter's chill descended upon their off-grid paradise, their heating system sprang into action, warming their home with gentle precision while minimizing energy waste.

But beyond the practical benefits, energy-efficient appliances were a symbol of their commitment to a more sustainable way of life. With every load of laundry and every meal prepared in their energy-efficient kitchen, they were reminded of the power of conscious consumption, of the profound impact that small choices can have on the health of our planet.

So, as you embark on your own off-grid journey, remember the importance of energy efficiency. Choose appliances that not only meet your needs but also respect the finite resources of our planet. For in the delicate dance of off-grid living, every watt saved is a step towards a brighter, more sustainable future for generations to come.

CHAPTER 7
SUSTAINABLE WASTE MANAGEMENT

In the tapestry of off-grid living, sustainable waste management is not just a practical necessity—it's a sacred duty. As our family settled into their new home amidst the rugged beauty of the wilderness, they quickly realized the importance of treading lightly upon the Earth, leaving behind only footprints of gratitude and respect.

Imagine their dismay as they gazed upon the pristine landscape marred by the unsightly scars of human waste—an all-too-common sight in today's throwaway society. Determined to do their part in preserving the natural beauty that surrounded them, our family embarked on a journey towards sustainable waste management, seeking to minimize their ecological footprint and maximize their stewardship of the land.

At the heart of their waste management strategy was the humble composting toilet—a marvel of simplicity and efficiency that transformed human waste into precious organic matter, returning it to the earth from whence it came. With each flush, they nourished the soil, replenishing its nutrients and fostering new life in a beautiful cycle of regeneration.

But composting toilets were just the beginning. Our family also embraced the principles of greywater recycling, finding creative ways to reuse water from showers, sinks, and laundry machines for irrigation and other non-potable purposes. Through careful design and thoughtful planning, they minimized their water usage and maximized their conservation efforts, ensuring that every drop was cherished and utilized to its fullest potential.

Yet perhaps the most profound aspect of their sustainable waste management journey was their commitment to mindful consumption and responsible disposal. From practicing the art of composting to reducing their reliance on single-use plastics, our family sought to live in harmony with the Earth, treating each resource with reverence and gratitude.

As they gazed upon their transformed landscape—a thriving oasis of life and vitality—they knew that their efforts had not been in vain. For in the delicate balance of waste and regeneration, they had discovered a profound truth—that by stewarding the land with care and intention, we can forge a brighter, more sustainable future for generations to come.

So, as you embark on your own journey towards sustainable waste management, remember the lessons of our intrepid family—embrace the principles of conservation, stewardship, and mindfulness, and let your actions speak louder than words. For in the tapestry of off-grid living, every choice we make has the power to shape the world around us, leaving behind a legacy of resilience, abundance, and beauty for all to enjoy.

Composting Toilets: Types and Construction Options

In the realm of off-grid living, one of the most essential yet often overlooked aspects of sustainability is waste management. Conventional flush toilets, reliant on water and sewage systems, are simply not feasible in off-grid settings. Enter composting toilets—a revolutionary solution that not only addresses the challenge of waste disposal but also transforms human waste into a valuable resource for the homestead.

Imagine a humble cabin nestled deep in the heart of the wilderness, where every drop of water is precious and every resource is utilized to its fullest potential. Here, composting toilets reign supreme, offering a simple yet ingenious way to turn waste into wealth. From simple DIY setups to sophisticated commercial systems, composting toilets come in various types and construction options, each tailored to suit the unique needs and preferences of off-grid dwellers.

Bucket or Basic Composting Toilets:
- In its simplest form, a bucket composting toilet consists of little more than a bucket, a seat, and a cover material such as sawdust or peat moss. Human waste is deposited into the bucket, where it is mixed with the cover material to promote decomposition and minimize Odors. Once the bucket is full, it is emptied into a designated composting bin or pile, where the waste undergoes natural decomposition over time.
- These basic composting toilets are inexpensive, easy to construct, and require minimal maintenance. They are well-suited for off-grid cabins, tiny homes, and remote campsites where simplicity and affordability are paramount.

Self-Contained Composting Toilets:
- Self-contained composting toilets are a step up from basic bucket toilets, offering a more refined and user-friendly experience. These toilets feature a built-in composting chamber where waste is mixed with bulking agents and undergoes aerobic decomposition.

- These toilets are compact, odorless, and require minimal installation—making them ideal for small dwellings and mobile homes. They are also equipped with ventilation systems to promote airflow and aid in the composting process, further reducing Odors and ensuring sanitary conditions.

Central Composting Systems:

- For larger off-grid communities or homesteads with multiple residents, central composting systems offer a scalable and efficient solution to waste management. These systems consist of a network of composting toilets connected to a centralized composting facility.
- Waste from individual toilets is collected and transported to the composting facility, where it is processed on a larger scale. These systems are often equipped with advanced features such as automatic mixing and aeration systems, ensuring optimal conditions for composting and minimizing manual labour.
- While central composting systems require more planning and investment upfront, they offer long-term benefits in terms of efficiency, convenience, and environmental sustainability.

DIY Composting Toilet Designs:

- One of the hallmarks of off-grid living is the spirit of DIY ingenuity, and composting toilets are no exception. From repurposed materials to custom-built designs, DIY enthusiasts have devised a myriad of creative solutions to meet their waste management needs.
- Whether it's converting an old shipping container into a composting toilet facility or fashioning a rustic outhouse using reclaimed lumber, the possibilities are limited only by one's imagination. DIY composting toilet designs are as diverse as the individuals who build them, reflecting a deep commitment to self-sufficiency and environmental stewardship.

In the world of off-grid living, composting toilets are more than just a practical necessity—they are a symbol of sustainability, resilience, and respect for the Earth. By embracing these innovative waste management solutions, off-grid dwellers can minimize their environmental footprint, reclaim valuable resources, and forge a deeper connection to the land they call home.

Greywater Systems for Non-Drinking Purposes

Greywater systems play a pivotal role in off-grid living, offering a sustainable solution for managing household water usage. In a world where water conservation is paramount, these systems provide a practical means of repurposing wastewater from sinks, showers, and washing machines for non-drinking purposes such as irrigation, toilet flushing, and outdoor cleaning.

Imagine a typical day in the life of our off-grid family. As they go about their daily routines, water flows freely from faucets and showers, serving its intended purposes before being whisked away down the drain. But instead of simply disappearing into the depths of the earth, this wastewater is captured, filtered, and redirected to a greywater system, where it undergoes a process of purification and repurposing.

At the heart of the greywater system lies a network of pipes and filtration mechanisms designed to remove impurities and contaminants, ensuring that the water is safe for its intended non-potable uses. From simple filtration screens to more advanced treatment processes such as biological filtration and UV sterilization, greywater systems come in a variety of configurations to suit different needs and preferences.

But perhaps the most remarkable aspect of greywater systems is their versatility. Whether you're watering a thriving vegetable garden, flushing toilets, or washing outdoor surfaces, greywater provides a valuable resource that can significantly reduce your reliance on freshwater sources. It's a simple yet ingenious solution that not only conserves water but also minimizes the environmental impact of traditional wastewater disposal methods.

Yet, like any system, greywater systems require careful planning and maintenance to ensure optimal performance and safety. Proper installation, regular inspection, and occasional maintenance are essential to prevent clogs, leaks, and microbial growth, safeguarding both the integrity of the system and the health of its users.

As our off-grid family discovered, greywater systems are more than just a practical solution—they're a testament to the power of innovation and resourcefulness in the face of adversity. They represent a tangible step towards a more sustainable future, where every drop of water is valued and cherished, and where our actions reflect a deep reverence for the precious gift of life.

So, as you embark on your own off-grid journey, consider incorporating a greywater system into your homestead. Not only will it help you conserve water and reduce your environmental footprint, but it will also serve as a powerful symbol of your commitment to living in harmony with the natural world.

Minimizing Waste Generation and Responsible Disposal

In the serene tranquillity of off-grid living, every resource is precious, every scrap of waste a potential opportunity. As our family settled into their new life, they quickly realized the importance of sustainable waste management—a cornerstone of self-sufficiency and environmental stewardship.

Minimizing waste generation became a guiding principle for our family, a mantra woven into the fabric of their daily lives. From the moment they awakened to the first rays of dawn to the quiet stillness of starlit nights, they were mindful of every action, every decision, and its impact on the delicate balance of their surroundings.

At the heart of their approach to waste management was a commitment to reducing consumption and embracing the ethos of "less is more." They carefully scrutinized every purchase, opting for products with minimal packaging and prioritizing durability and longevity over convenience and disposability. Through conscious consumption, they minimized the flow of waste into their lives, preserving precious resources and reducing their environmental footprint.

But even the most mindful consumer inevitably generates waste, and so our family turned their attention to responsible disposal—a task that demanded creativity, ingenuity, and a deep respect for the Earth. They embraced the principles of recycling and upcycling, transforming discarded materials into new treasures with a little creativity and elbow grease. From repurposing glass jars as storage containers to fashioning old clothing into quilts and rugs, they discovered the transformative power of resourcefulness, turning waste into wealth with every stitch and brushstroke.

Composting became another cornerstone of their waste management strategy, a natural solution to the age-old problem of organic waste. They established compost piles teeming with life, where food scraps and garden clippings were transformed into rich, nutrient-dense soil—a precious gift to nourish their beloved garden and sustain the circle of life.

Yet, perhaps the most profound lesson our family learned in their journey towards sustainable waste management was the importance of mindfulness and intentionality. They approached every task with a sense of reverence and gratitude, mindful of the interconnectedness of all living things and the profound impact of their actions on the world around them.

So, as you embark on your own journey towards self-sufficiency, remember the importance of sustainable waste management—a cornerstone of off-grid living and environmental stewardship. Embrace the principles of mindful consumption, responsible disposal, and creative resourcefulness, and let your actions speak volumes about your commitment to a brighter, more sustainable future for generations to come.

PART 4

ESSENTIAL SKILLS AND KNOWLEDGE FOR NO GRID LIVING

CHAPTER 8
TOOL MAINTENANCE AND REPAIR

In the rugged terrain of off-grid living, tools are not just instruments—they are lifelines, essential companions on the journey towards self-sufficiency. From the humble hammer to the mighty chainsaw, each tool plays a vital role in shaping our ability to thrive in the wilderness. But like any trusted ally, these tools require care and maintenance to ensure they remain in peak condition, ready to serve us when we need them most.

Imagine our family's dismay as they set out to build their off-grid

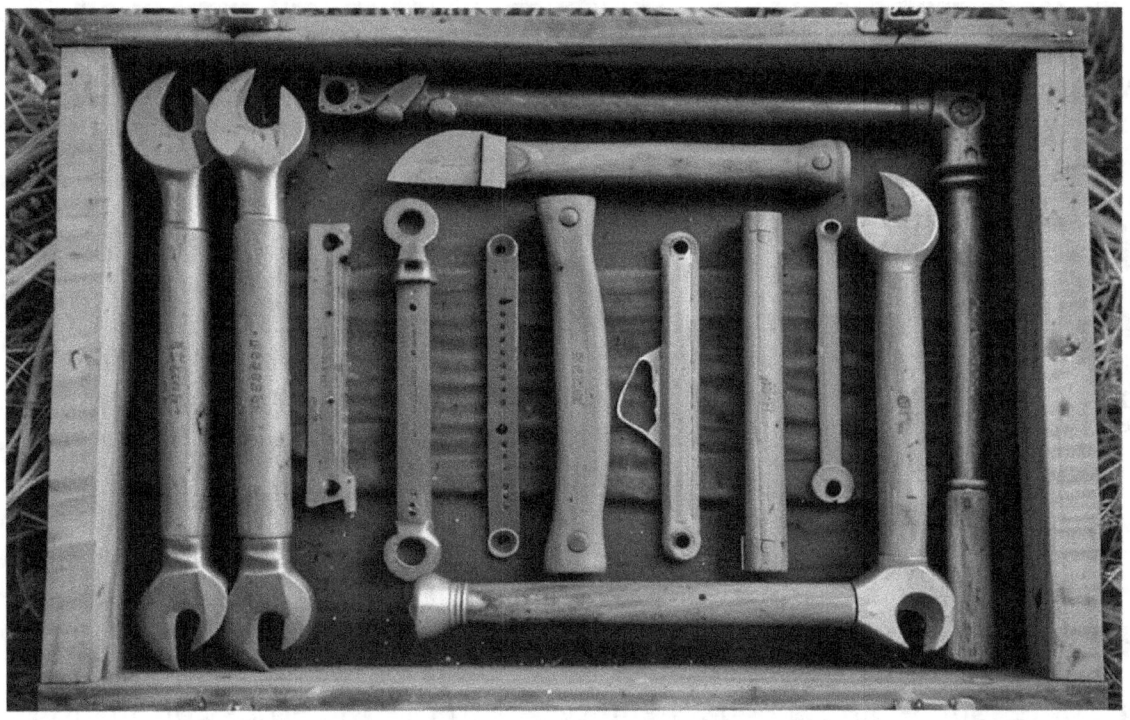

homestead, only to find their trusty tools rusted and worn from neglect. With a sinking heart, they realized that without proper maintenance, their tools were useless, mere relics of a bygone era. Determined to rectify their oversight, they set out on a quest to learn the art of tool maintenance and repair, transforming their rusty relics into shining symbols of resilience and resourcefulness.

At the heart of tool maintenance lies a simple truth: prevention is better than cure. By implementing a regular maintenance routine, our family learned to keep their tools in tip-top shape, preventing rust, corrosion, and wear before they could take hold. From cleaning and lubricating to sharpening and calibrating, they discovered the importance of regular upkeep in preserving the longevity and performance of their tools.

But even the most well-maintained tools are not immune to the ravages of time and use. As our family soon discovered, breakdowns and malfunctions are an inevitable part of off-grid living, requiring swift and decisive action to resolve. Armed with a newfound understanding of tool repair, they learned to diagnose and troubleshoot common issues, from jammed saw blades to faulty electrical connections, with confidence and skill.

Yet, perhaps the most valuable lesson our family learned was the importance of resourcefulness in the face of adversity. In the absence of ready-made replacements, they discovered the art of improvisation, repurposing materials and devising ingenious solutions to keep their tools in working order. From fashioning makeshift handles out of fallen branches to fashioning replacement parts from salvaged metal scraps, they embraced the challenge of DIY repair with creativity and determination.

As our family's off-grid homestead flourished and grew, so too did their appreciation for the humble tools that had accompanied them on their journey. Through the trials and tribulations of tool maintenance and repair, they forged a deep bond with their instruments, recognizing them not just as objects, but as partners in their quest for self-sufficiency.

So, as you embark on your own off-grid adventure, remember the lessons of Chapter 8: Tool Maintenance and Repair. Treat your tools with care and respect, and they will reward you with years of faithful service. Learn to diagnose and troubleshoot common issues, and you'll never be caught off guard by a sudden breakdown. And above all, embrace the spirit of resourcefulness and ingenuity, for in the wilderness of off-grid living, there are no problems—only opportunities to learn, grow, and thrive.

Essential Tools for Off-Grid Living

n the rugged terrain of off-grid living, where self-sufficiency is the name of the game, having the right tools at your disposal can mean the difference between thriving and merely surviving. As our intrepid family discovered on their journey towards self-sufficiency, the right tools are not just instruments of labour—they're indispensable allies, empowering you to tackle any challenge that comes your way with confidence and ease.

Imagine their delight as they unpacked their toolkit for the first time, each tool a promise of possibility, a testament to their determination to carve out a life of independence and resilience. From sturdy axes and reliable saws to versatile multi-tools and precision hand drills, their arsenal was as diverse as it was essential, each tool serving a unique purpose in their quest for off-grid mastery.

But while the right tools are crucial, knowing how to use them effectively is equally important. That's why our family wasted no time in honing their skills, mastering the art of tool maintenance and repair with unwavering dedication and precision. From sharpening blades and oiling hinges to replacing worn-out parts and troubleshooting mechanical issues, they approached each task with meticulous attention to detail, ensuring that their tools remained in peak condition at all times.

But perhaps the most valuable tool of all was not found in their toolkit, but within themselves—their unwavering spirit of ingenuity, their unyielding determination to overcome any obstacle in their path. Armed with this inner strength, they transformed even the most rudimentary of tools into instruments of ingenuity, finding creative solutions to the myriad challenges of off-grid living with grace and aplomb.

So, as you embark on your own off-grid adventure, remember that while tools are important, it's your resourcefulness, determination, and ingenuity that will ultimately carry you through. With the right tools at your disposal and the right mindset in your heart, there's no challenge too great, no obstacle too daunting, and no dream too ambitious for you to conquer. So, arm yourself with the tools of knowledge, the tools of determination, and the tools of resilience, and embark on your journey towards self-sufficiency with confidence, courage, and grace.

Basic Repairs and Maintenance for Common Tools

In the rugged terrain of off-grid living, tools are more than just instruments—they're lifelines, essential companions in the daily struggle to build, create, and thrive. But in the harsh embrace of nature, even the sturdiest tools can succumb to wear and tear, requiring diligent maintenance and occasional repairs to keep them in peak condition.

As our family settled into their off-grid homestead, they quickly learned the importance of basic repairs and maintenance for their beloved tools. From the trusty axe that split firewood to the humble hammer that pounded nails into place, each tool played a vital role in their quest for self-sufficiency, demanding care and attention in return.

Imagine their frustration as they struggled to wield a dull saw, its teeth dulled by months of hard use. Or their dismay as they discovered a leak in their trusty watering can, its once-reliable seal compromised by the ravages of time. Yet, with each challenge came an opportunity—to learn, to grow, and to master the art of tool maintenance and repair.

At the heart of tool maintenance lies a simple truth: prevention is key. By regularly inspecting and cleaning your tools, you can nip potential problems in the bud before they escalate into full-blown crises. A quick wipe-down with a rag and a dab of oil can work wonders, keeping rust at bay and ensuring smooth operation for years to come.

But sometimes, despite your best efforts, tools inevitably succumb to the rigors of off-grid life, requiring more extensive repairs to restore them to their former glory. From sharpening a dull blade to replacing a broken handle, these repairs demand patience, precision, and a steady hand.

Imagine the satisfaction our family felt as they breathed new life into their worn-out tools, their efforts rewarded with renewed Vigor and efficiency. With each repair, they grew more adept, more confident in their ability to conquer any challenge that came their way.

But perhaps the most valuable lesson they learned was the importance of resilience—the ability to adapt and innovate in the face of adversity. In the wild embrace of off-grid living, tools are more than just instruments—they're symbols of our ingenuity, our resourcefulness, and our unyielding determination to thrive against all odds.

So, as you embark on your own off-grid adventure, remember the lessons of tool maintenance and repair. Treat your tools with care and respect, and they will serve you faithfully in return. And when the inevitable challenges arise, embrace them as opportunities—to learn, to grow, and to master the art of self-sufficiency in the wild embrace of nature.

Troubleshooting and Problem-Solving Techniques

In the rugged terrain of off-grid living, challenges are inevitable. Whether it's a malfunctioning solar panel, a leaky rainwater collection system, or a pest infestation in your garden, problems are bound to arise from time to time. But fear not, for with the right troubleshooting and problem-solving techniques at your disposal, you can tackle any obstacle that comes your way with confidence and ingenuity.

1. Identify the Problem: The first step in solving any problem is identifying its root cause. Take the time to carefully observe and assess the situation, paying close attention to any symptoms or warning signs that may offer clues to the underlying issue.

2. Gather Information: Once you've identified the problem, gather as much information as possible to help you better understand its nature and scope. Consult user manuals, online resources, or seek advice from fellow off-grid enthusiasts to gain valuable insights and perspectives.

3. Break it Down: Break the problem down into smaller, more manageable components. This will help you to isolate the issue and identify potential areas of concern more effectively, making it easier to devise a targeted solution.

4. Brainstorm Solutions: Get creative and brainstorm potential solutions to the problem at hand. Consider all possible options, no matter how unconventional they may seem, and weigh the pros and cons of each approach before making a decision.

5. Implement a Plan of Action: Once you've identified a viable solution, put your plan into action with confidence and determination. Be prepared to adapt and iterate as needed, and don't be afraid to seek help or advice from others if you encounter unexpected challenges along the way.

6. Test and Evaluate: After implementing your solution, test its effectiveness and evaluate the results. Did it solve the problem? If not, don't be discouraged—use this opportunity to learn from your mistakes and refine your approach for future troubleshooting endeavours.

7. Learn from the Experience: Every problem presents an opportunity for growth and learning. Take the time to reflect on your troubleshooting process, noting what worked well and what could be improved upon next time. Armed with this newfound knowledge, you'll be better equipped to tackle future challenges with confidence and resilience.

8. Stay Positive and Persistent: Finally, remember to stay positive and persistent in the face of adversity. Off-grid living is a journey filled with ups and downs, but with the right mindset and a willingness to persevere, you can overcome any obstacle that stands in your way and emerge stronger and more resilient on the other side.

By mastering the art of troubleshooting and problem-solving, you'll not only become a more effective off-grid homesteader, but you'll also cultivate a sense of empowerment and self-reliance that will serve you well on your journey towards self-sufficiency. So embrace the challenges that come your way, armed with the knowledge and skills to tackle them head-on, and watch as you transform obstacles into opportunities and setbacks into successes in the wild and wondrous world of off-grid living.

CHAPTER 9
FIRST AID AND EMERGENCY PREPAREDNESS

In the wilderness, where the nearest hospital may be hours away and help is not just a phone call away, mastering first aid and emergency preparedness isn't just a skill—it's a matter of life and death. As our intrepid family ventured deeper into the heart of off-grid living, they quickly realized the critical importance of being able to respond effectively to medical emergencies and unforeseen disasters.

Imagine the scene: A peaceful afternoon in the forest is shattered by the sound of a loud crash, followed by the panicked cries of a family member. Racing to the scene, our family discovers their loved one lying injured on the ground, blood staining the earth beneath them. In that moment of crisis, every second counts, and the knowledge and skills they've acquired in first aid training become their most valuable assets.

First aid isn't just about applying bandages and administering CPR—it's about remaining calm under pressure, assessing the situation quickly and accurately, and taking decisive action to stabilize the patient until help arrives. For our family, first aid training was more than just a precaution; it was a lifeline in moments of crisis, allowing them to respond effectively to everything from minor cuts and bruises to serious injuries and medical emergencies.

But first aid is just one piece of the puzzle when it comes to emergency preparedness. In the wilderness, where storms can rage and disasters can strike without warning, being prepared for the unexpected is essential. From natural disasters like wildfires and floods to medical emergencies like heart attacks and allergic reactions, our family learned to anticipate the unexpected and take proactive steps to protect themselves and their loved ones.

Imagine their relief as they emerge unscathed from a devastating storm, thanks to the emergency supplies and evacuation plan they put in place beforehand. Or the sense of empowerment as they administer life-saving first aid to a neighbour in need, their hands steady and their minds clear under pressure. In the wilderness, where every day brings new challenges and uncertainties, being prepared isn't just a luxury—it's a necessity.

So, as you embark on your own off-grid adventure, remember the critical importance of first aid and emergency preparedness. Arm yourself with the knowledge and skills you need to respond effectively to medical emergencies and unforeseen disasters, and take proactive steps to protect yourself and your loved ones from harm. With the right training, mindset, and preparation, you can navigate the challenges of off-grid living with confidence and grace, knowing that you're prepared for whatever the wilderness may throw your way.

Building a First-Aid Kit for Off-Grid Situations

In the rugged terrain of off-grid living, where the nearest hospital may be hours away and professional medical help is not readily available, a well-equipped first-aid kit can mean the difference between life and death. As our intrepid family ventured deeper into the wilderness, they quickly learned the importance of being prepared for any medical emergency that might arise.

Imagine their relief as they unpacked their carefully curated first-aid kit, its contents neatly organized and ready for action. From bandages and antiseptic wipes to splints and pain relievers, their kit was stocked with everything they might need to handle minor injuries and ailments with ease. But more than just a collection of supplies, their first-aid kit was a lifeline—a symbol of their commitment to safety, preparedness, and the well-being of their loved ones.

But building a first-aid kit for off-grid situations is no simple task. It requires careful planning, foresight, and a thorough understanding of the unique challenges posed by remote living. Fortunately, our family was up to the challenge, drawing on their own experiences and the advice of seasoned off-grid veterans to create a kit that was both comprehensive and practical.

First and foremost, their kit included basic wound care supplies such as sterile gauze pads, adhesive bandages, and medical tape. These items were essential for treating cuts, scrapes, and other minor injuries that might occur during everyday activities around the homestead.

In addition to wound care supplies, their kit also contained a variety of over-the-counter medications to address common ailments such as headaches, fevers, and allergies. Pain relievers, antihistamines, and anti-diarrheal medications were all included, ensuring that they could manage a wide range of medical issues without the need for a trip to the pharmacy.

But perhaps the most important aspect of their first-aid kit was its versatility. Recognizing that off-grid living often involves activities that carry a higher risk of injury, such as chopping firewood or hiking in remote areas, they made sure to include supplies for more serious medical emergencies as well. Splints, tourniquets, and CPR masks were all part of their kit, allowing them to respond effectively to everything from broken bones to cardiac arrest.

Yet, even the most well-equipped first-aid kit is only as good as the person using it. That's why our family made sure to familiarize themselves with the contents of their kit and undergo basic first-aid training before setting out on their off-grid adventure. They practiced treating common injuries and rehearsed emergency scenarios, ensuring that they would be calm, confident, and competent in the event of a medical crisis.

As they journeyed deeper into the wilderness, our family took comfort in knowing that they were prepared for whatever challenges might come their way. Their first-aid kit was more than just a collection of supplies—it was a symbol of their resilience, their resourcefulness, and their unwavering commitment to the safety and well-being of their family. And as they faced the trials and tribulations of off-grid living head-on, they did so with the confidence that comes from knowing that they were ready for anything.

Basic First Aid Skills and Response Protocols

In the rugged terrain of off-grid living, where every day brings new challenges and adventures, being prepared for medical emergencies is not just prudent—it's essential. As our family settled into their remote homestead, they quickly realized the importance of equipping themselves with basic first aid skills and developing robust response protocols to handle any situation that might arise.

Imagine the scene: A peaceful afternoon in the wilderness is shattered by a sudden cry for help. With hearts pounding and adrenaline coursing through their veins, our family rushes to the scene, where they find their neighbour, clutching his arm in agony. Without hesitation, they spring into action, drawing on their training to stabilize the situation and provide critical care until help arrives.

Basic first aid skills are the foundation of any emergency response plan, empowering individuals to assess, treat, and manage a wide range of injuries and medical conditions. From cuts and bruises to fractures and burns, knowing how to administer first aid can mean the difference between life and death in a remote setting where professional medical help may be hours or even days away.

But first aid is more than just bandaging wounds and applying splints—it's about remaining calm under pressure, making quick decisions, and providing compassionate care to those in need. It's about knowing when to seek help and when to take decisive action, even in the face of uncertainty and fear.

For our family, basic first aid training became a cornerstone of their off-grid lifestyle, instilling them with the confidence and competence to handle medical emergencies with grace and skill. From CPR and wound care to immobilization and evacuation techniques, they honed their skills through hands-on training and real-world experience, ensuring that they were always ready to spring into action when disaster struck.

But perhaps the most valuable lesson they learned was the importance of preparation and prevention. By conducting regular safety assessments, stocking their first aid kit with essential supplies, and practicing emergency response drills, they were able to mitigate risks and minimize the likelihood of accidents and injuries on their homestead.

So, as you embark on your own off-grid adventure, remember the importance of basic first aid skills and response protocols. Whether you're a seasoned wilderness enthusiast or a novice homesteader, investing the time and effort to develop these essential skills could one day save a life—and in the remote wilderness of off-grid living, that's a priceless gift indeed.

Emergency Preparedness Planning

Finally, imagine this: A powerful earthquake strikes, leaving our family's off-grid homestead in shambles. With no power, no communication, and limited resources, they must rely on their emergency preparedness plan to weather the storm and emerge unscathed.

Emergency preparedness planning is the cornerstone of resilience in the face of disaster, providing you with a roadmap for navigating even the direst situations. Start by identifying potential risks and hazards in your area—earthquakes, wildfires, floods, and severe weather—and develop a comprehensive plan for how to respond to each scenario.

This plan should include:

- Evacuation routes and meeting points
- Contact information for emergency services and local authorities

- Supplies and resources for sheltering in place or evacuating
- Protocols for communication and staying informed during a crisis
- Backup power sources and alternative forms of communication (such as a hand-crank radio or satellite phone)
- And most importantly, practice and review your emergency plan regularly to ensure that you and your family are prepared to act quickly and decisively when disaster strikes.

In the end, first aid and emergency preparedness are not just skills to be learned—they are a mindset, a way of life that empowers us to face the unknown with courage and resilience. So as we embark on this journey into the heart of no grid living, let us remember the importance of being prepared, of equipping ourselves with the tools and knowledge to handle whatever challenges may come our way.

CHAPTER 10
SECURING YOUR OFF-GRID PROPERTY

In the remote expanse of the wilderness, where the echoes of civilization fade into the whispers of the wind, lies a sanctuary of self-sufficiency—a place where our intrepid family found refuge from the chaos of the modern world. But amidst the tranquillity of their off-grid haven, they soon realized that security was paramount—a shield against the uncertainties that lurked beyond the safety of their homestead.

Imagine their sense of urgency as they fortified their off-grid property, transforming it into a fortress of resilience—a bastion of safety in a world fraught with danger. From perimeter security to safeguarding their precious resources, they spared no expense in their quest to protect their newfound way of life.

At the heart of their security strategy lay the importance of perimeter security—an invisible barrier that stood between them and the outside world. With razor wire and motion sensors, they encircled their property, warding off trespassers and would-be intruders with a silent warning. But security wasn't just about keeping others out—it was also about protecting what lay within.

With meticulous care, they secured their food and water supplies, fortifying their stores against the ravages of nature and the prying eyes of opportunistic predators. From reinforced storage containers to hidden caches, they ensured that their provisions remained safe and secure, ready to sustain them through even the harshest of times.

But security wasn't just a physical barrier—it was also a state of mind. Through community vigilance and mutual support, they forged bonds of trust and camaraderie with their fellow off-gridders, creating a network of resilience that stretched far beyond the boundaries of their own property. Together, they watched over each other, sharing resources and knowledge, and standing united against the challenges that threatened to tear them apart.

As you embark on your own off-grid journey, remember that security is not just a luxury—it's a necessity. Whether you're safeguarding your property against intruders or protecting your resources against the ravages of nature, security is the foundation upon which self-sufficiency is built. So, fortify your defences, forge bonds of trust, and embrace the peace of mind that comes from knowing that you are prepared for whatever challenges lie ahead.

Perimeter Security and Deterrents for Trespassers

As our intrepid family settled into their off-grid homestead, they soon realized that security was paramount in their quest for self-sufficiency. With the vast expanse of wilderness surrounding their property, they knew that protecting their land from trespassers and intruders would be essential to safeguarding their newfound way of life.

Perimeter security became their first line of defence—a formidable barrier against the uncertainties of the wild. But for our family, security wasn't just about fortifying their property with fences and locks; it was about fostering a sense of safety and peace of mind in their remote corner of the world.

Their journey towards securing their off-grid property began with a thorough assessment of their land's vulnerabilities. They mapped out potential entry points and identified areas of concern, from dense forest cover to hidden trails frequented by wildlife and humans alike. Armed with this knowledge, they set to work implementing a multi-layered approach to perimeter security, combining physical barriers with strategic deterrents to keep trespassers at bay.

Fencing became their first line of defence, encircling their property with sturdy barriers designed to deter both human intruders and curious wildlife. But our family knew that fences alone wouldn't be enough to keep determined trespassers at bay. They supplemented their physical barriers with a variety of deterrents, from motion-activated lights and alarms to strategically placed signage warning of the consequences of trespassing.

Yet, perhaps the most effective deterrent of all was the sense of community they fostered amongst their neighbours. Through mutual support and vigilance, they formed a tight-knit network of like-minded individuals dedicated to protecting each other's properties from harm. Together, they patrolled their shared boundaries, keeping a watchful eye out for any signs of trouble and standing united against any threat to their way of life.

But securing their off-grid property wasn't just about keeping intruders out—it was also about creating a sense of sanctuary and serenity in their remote corner of the world. They cultivated a welcoming atmosphere, inviting visitors to respect their land and tread lightly upon it. And through their efforts, they transformed their off-grid homestead into a haven of peace and tranquillity, where the rhythms of nature reign supreme and the spirit of community thrives.

So, as you embark on your own journey towards securing your off-grid property, remember that perimeter security is not just about building walls—it's about building relationships, fostering a sense of belonging, and creating a sanctuary where you can thrive in harmony with the land. With vigilance, determination, and a dash of ingenuity, you too can safeguard your off-grid paradise and embrace a life of true self-sufficiency.

Securing Your Food and Resources

In the vast expanse of off-grid living, securing your food and resources is not just a practical necessity—it's a sacred duty, a vital lifeline that sustains you through the ebb and flow of the seasons, the trials and triumphs of self-sufficiency. As our intrepid family learned on their journey towards off-grid resilience, the key to securing their sustenance lay not just in stockpiling provisions, but in cultivating a deep respect for the land, a profound understanding of its rhythms and cycles, and a fierce determination to protect and preserve its bounty for generations to come.

Imagine their delight as they harvested their first fruits and vegetables from their off-grid garden, their hands stained with the rich earth of their newfound home. From plump tomatoes bursting with flavour to crisp cucumbers glistening with morning dew, each harvest was a testament to their hard work and dedication, a celebration of the abundance that lay within their reach when they dared to embrace the challenge of self-sufficiency.

But as the seasons changed and the bounty of summer gave way to the barren landscapes of winter, our family quickly realized the importance of planning ahead and diversifying their food sources. They turned to the age-old practice of food preservation, learning the art of canning, drying, and fermenting to ensure that their pantry remained well-stocked throughout the lean months ahead. From jars of homemade pickles to strings of sun-dried tomatoes hanging from the rafters, their larder became a veritable treasure trove of preserved delights, a testament to their ingenuity and resourcefulness in the face of scarcity.

Yet, securing their food and resources went beyond mere sustenance—it was also a matter of safeguarding their livelihood and protecting their homestead from the ever-present threat of predators and pests. They fortified their garden with sturdy fences and watchful scarecrows, standing guard against marauding deer and cunning rabbits intent on plundering their hard-earned harvest. They installed motion-activated lights and alarms to deter would-be intruders, ensuring that their sanctuary remained safe and secure even in the dead of night.

But perhaps the most crucial aspect of securing their food and resources was fostering a sense of community and cooperation with their fellow off-gridders. They traded surplus produce with neighbouring homesteads, sharing seeds, knowledge, and camaraderie in a spirit of mutual support and solidarity. Together, they formed a tight-knit network of resilience, a web of interconnectedness that stretched across the vast expanse of the wilderness, binding them together in a shared mission to thrive in harmony with the land.

So, as you embark on your own journey towards off-grid living, remember that securing your food and resources is not just a practical necessity—it's a sacred pact, a covenant with the earth and with each other to honour and protect the precious gifts of nature that sustain us all. With determination, ingenuity, and a generous spirit of cooperation, you too can forge a path towards self-sufficiency, securing your sustenance and your future in the wild embrace of off-grid living.

Off-Grid Communication Options and Emergency Alerts

In the remote wilderness, far from the reach of traditional communication networks, staying connected and informed can be a matter of life and death. As our intrepid family discovered, off-grid communication options and emergency alerts are essential tools for navigating the challenges of self-sufficiency and ensuring the safety and well-being of all who call the wild embrace of nature their home.

Imagine the scene: a sudden storm descends upon the valley, its fury unleashing torrents of rain and howling winds that threaten to uproot trees and flood rivers. In the midst of this chaos, our family finds themselves cut off from the outside world, their only lifeline to civilization severed by the wrath of Mother Nature.

But fear not, for in the face of adversity, our family is armed with a suite of off-grid communication options and emergency alerts designed to keep them informed and connected even in the direst of circumstances. From ham radio communication to satellite internet options, they harness the power of technology to bridge the gap between their remote wilderness retreat and the outside world.

Ham radio communication emerges as a stalwart ally in their battle against isolation, offering reliable long-range communication that transcends the limitations of traditional cell networks. With a ham radio license in hand and a trusty radio set at their side, our family is able to broadcast vital updates, coordinate rescue efforts, and stay connected with fellow off-grid pioneers and emergency responders alike.

But technology alone cannot guarantee safety in the wilderness. In the event of a natural disaster or medical emergency, timely alerts and notifications are crucial for ensuring a swift and effective response. That's where emergency alert systems come into play, providing real-time updates on weather warnings, evacuation orders, and other critical information that can mean the difference between life and death.

Whether it's a handheld weather radio tuned to NOAA's National Weather Service or a smartphone equipped with emergency alert apps like FEMA's Wireless Emergency Alerts (WEA) system, our family is never caught off guard by sudden changes in the weather or unforeseen emergencies. With their off-grid communication options and emergency alerts at the ready, they stand poised to weather any storm and emerge stronger and more resilient than ever before.

So, as you embark on your own off-grid adventure, remember the importance of staying connected and informed in the wilderness. Arm yourself with the tools and technology you need to communicate with the outside world and receive timely alerts and notifications when disaster strikes. With a little preparation and foresight, you too can navigate the challenges of off-grid living with confidence and grace, ensuring your safety and security in even the most remote corners of the earth.

Part 5

Advanced Topics and Considerations

CHAPTER 11
OFF-GRID COMMUNICATION

In the vast expanse of off-grid living, communication isn't just a convenience—it's a lifeline, connecting us to the outside world and ensuring our safety and security in even the most remote corners of the wilderness. As our intrepid family discovered, mastering the art of off-grid communication is essential for navigating the challenges of self-sufficiency with confidence and grace.

Imagine their excitement as they embarked on their off-grid journey, leaving behind the hustle and bustle of city life for the quiet solitude of the wilderness. Yet, as they settled into their new home, they soon realized that staying connected wasn't as simple as flipping a switch or picking up a phone. In the absence of traditional communication infrastructure, they were forced to get creative, exploring a myriad of innovative solutions to stay in touch with the world beyond.

At the heart of their off-grid communication strategy was the venerable ham radio—a time-tested technology beloved by amateur radio enthusiasts the world over. With its long-range capabilities and ability to operate independently of traditional infrastructure, the ham radio became their lifeline to the outside world, allowing them to communicate with fellow off-gridders, access emergency services, and stay informed about local events and weather conditions.

But the ham radio was just the beginning. As our family delved deeper into the world of off-grid communication, they discovered a wealth of alternative technologies and techniques, each one offering unique advantages and capabilities. From satellite phones and portable two-way radios to low-tech signalling devices like signal mirrors and smoke signals, they quickly learned that staying connected off the grid was as much about creativity and resourcefulness as it was about technology.

Yet, perhaps the most powerful form of off-grid communication was the simplest—the art of storytelling. As they gathered around the campfire each night, our family shared tales of their adventures and misadventures, weaving a tapestry of shared experiences that bound them together as a community. In the absence of traditional communication infrastructure, they found solace in the timeless ritual of storytelling, using it to bridge the gap between past and present, between isolation and connection.

So, as you embark on your own off-grid adventure, remember that communication isn't just about technology—it's about connection, community, and the timeless art of storytelling. Whether you're chatting with fellow off-gridders on the ham radio or sharing stories around the campfire, embrace the power of communication to forge bonds that transcend distance and time, and to navigate the challenges of off-grid living with courage and grace.

Ham Radio Communication and Licensing

In the vast expanse of off-grid living, communication is not just a convenience—it's a lifeline, connecting us to the outside world and providing vital information and support when we need it most. And when it comes to reliable, long-distance communication in remote areas, few technologies rival the time-tested reliability of ham radio.

Imagine the scene: A fierce storm rages outside, cutting off all conventional forms of communication and leaving our intrepid family stranded in the wilderness. With no phone signal and no internet access, they turn to their trusty ham radio—a beacon of hope in the midst of the storm. With practiced hands, they tune their radio to the designated frequency and send out a distress call, praying for a response.

Ham radio, also known as amateur radio, has long been a staple of emergency communication in remote areas and disaster zones. Unlike conventional communication technologies, which rely on centralized infrastructure and can be easily disrupted by natural disasters or technical failures, ham radio operates independently of external networks, using radio waves to transmit messages over long distances.

But ham radio is more than just a tool for emergency communication—it's a vibrant community of enthusiasts, hobbyists, and emergency responders united by a shared passion for radio technology. From building DIY antennas to participating in international contests and competitions, ham radio offers endless opportunities for learning, exploration, and camaraderie.

Yet, to unlock the full potential of ham radio, one must first obtain the necessary licensing. In many countries, ham radio operators are required to pass a licensing exam administered by the government or a designated authority, demonstrating their proficiency in radio theory, regulations, and operating procedures.

But fear not, for the rewards of becoming a licensed ham radio operator far outweigh the challenges. With your license in hand, you gain access to a world of possibilities, from chatting with fellow enthusiasts on local repeaters to participating in global networks like the Amateur Radio Emergency Service (ARES) and the Radio Amateur Civil Emergency Service (RACES).

So, whether you're a seasoned radio aficionado or a newcomer eager to explore the world of amateur radio, obtaining your ham radio license is the first step towards unlocking the full potential of this powerful communication technology. With dedication, study, and a healthy dose of curiosity, you too can join the ranks of licensed ham radio operators and embark on a journey of discovery, connection, and community that spans the globe.

Satellite Internet Options

In the vast expanse of off-grid living, access to reliable internet can sometimes feel like a luxury beyond reach. Yet, in today's interconnected world, staying connected is more important than ever, whether it's for remote work, online education, or simply staying in touch with loved ones. That's where satellite internet options come into play, offering a lifeline to those living off the grid in remote or rural areas.

Imagine our family, nestled deep in the heart of the wilderness, their cabin perched atop a sun-kissed hilltop with panoramic views of the surrounding landscape. While their secluded paradise offers respite from the hustle and bustle of modern life, it also presents challenges, particularly when it comes to staying connected to the outside world.

Enter satellite internet—their gateway to the digital realm, a high-speed connection to the vast expanse of cyberspace, no matter how remote their location may be. With a satellite dish installed on their property, our family gains access to fast and reliable internet service, allowing them to browse the web, send emails, and even stream movies and music from the comfort of their off-grid oasis.

But satellite internet options offer more than just connectivity; they also provide a sense of security and peace of mind, knowing that help is just a click away in case of emergencies. Whether it's accessing weather forecasts, communicating with emergency services, or staying informed about local news and events, satellite internet ensures that our family remains connected to the outside world, even as they embrace the serenity of off-grid living.

Of course, satellite internet options are not without their challenges. From high installation costs to data caps and latency issues, there are drawbacks to consider. Yet, for many off-grid residents, the benefits far outweigh the drawbacks, offering a lifeline to the digital world and a vital connection to the broader community.

So, as you embark on your own off-grid journey, consider satellite internet options as a valuable tool in your toolkit, offering connectivity, convenience, and peace of mind in an increasingly interconnected world. With satellite internet by your side, you can stay connected, stay informed, and stay inspired, no matter how far off the grid you may roam.

Low-Tech Communication Methods

In the remote expanses of off-grid living, where the whispers of nature drown out the cacophony of modern life, communication takes on a new significance. While smartphones and high-speed internet may be out of reach, our intrepid family discovered a world of low-tech communication methods that allowed them to stay connected with loved ones and their community, even in the most remote corners of the wilderness.

Imagine the thrill of receiving a handwritten letter from a distant friend, carried by the gentle hand of a passing traveller or nestled within the pages of a care package sent via snail mail. In an age of instant messaging and digital communication, the simple act of receiving a letter became a cherished ritual, a tangible reminder of the enduring bonds that connect us to one another.

But letters were just the beginning. Our family also embraced the timeless art of Morse code, tapping out messages with a simple telegraph key or using flashes of light to communicate across vast distances. With nothing more than a basic understanding of Morse code and a homemade signalling device, they were able to send messages of love, friendship, and solidarity to far-flung corners of the globe.

And let us not forget the humble carrier pigeon, whose swift wings and keen instincts made them invaluable messengers in times of need. With a trained flock of pigeons at their disposal, our family was able to send urgent messages across great distances, bypassing the limitations of modern communication technology and relying instead on the age-old wisdom of nature's own couriers.

But perhaps the most powerful form of low-tech communication was the simple act of gathering around a crackling campfire, sharing stories and laughter beneath the star-studded sky. In these moments of connection, words became unnecessary, replaced instead by the silent language of shared experience and mutual understanding.

So, as you embark on your own off-grid adventure, remember that communication is not just about technology—it's about connection. Whether you're sending a letter, tapping out Morse code, or gathering around a campfire with loved ones, cherish each moment of connection as a precious gift, a reminder of the enduring bonds that unite us all in the great tapestry of life.

CHAPTER 12
LEGAL AND REGULATORY CONSIDERATIONS

In the vast tapestry of off-grid living, legal and regulatory considerations serve as the threads that bind our aspirations for self-sufficiency with the realities of the modern world. As our intrepid family navigated the uncharted waters of off-grid living, they quickly discovered that while the wilderness may offer freedom and solace, it also comes with its fair share of rules and regulations—rules that must be understood and respected if one is to thrive in harmony with both nature and society.

Imagine their surprise as they uncovered the complex web of building permits and zoning regulations that govern the construction of off-grid structures. From the size and location of their cabin to the materials used in its construction, every aspect of their home was subject to scrutiny and oversight by local authorities. Yet, far from being a hindrance, these regulations served as a guiding light, helping our family to navigate the maze of bureaucracy and ensure that their dreams of self-sufficiency remained firmly rooted in legality and compliance.

But building permits were just the beginning. As our family delved deeper into the intricacies of off-grid living, they soon encountered a myriad of other legal considerations, from water rights and rainwater harvesting regulations to animal husbandry laws and waste disposal ordinances. Each one presented its own unique set of challenges and complexities, yet each also offered an opportunity for growth and learning, as our family sought to reconcile their desire for autonomy with the realities of communal living.

For our family, legal and regulatory compliance became not just a requirement, but a moral imperative—a reflection of their commitment to responsible stewardship of the land and respect for the communities in which they lived. From obtaining the necessary permits for their off-grid homestead to navigating the intricacies of water rights and animal husbandry regulations, they approached each challenge with humility and determination, seeking not just to comply with the law, but to embrace it as a guiding principle in their quest for self-sufficiency.

So, as you embark on your own journey towards off-grid living, remember that legal and regulatory considerations are not obstacles to be overcome, but opportunities to be embraced. With patience, diligence, and a willingness to engage with the complexities of the modern world, you too can navigate the legal landscape of off-grid living with confidence and grace, ensuring that your dreams of self-sufficiency remain firmly rooted in legality, responsibility, and respect.

Building Permits and Zoning Regulations for Off-Grid Structures

In the realm of off-grid living, navigating the labyrinth of building permits and zoning regulations can feel like traversing uncharted territory. Yet, understanding and complying with these regulations is crucial for ensuring the legality and safety of your off-grid structures. Let's delve deeper into this critical aspect of off-grid living.

Understanding Building Permits:

Building permits are government-issued documents that grant legal authorization to undertake construction or renovation projects on a property. They serve to ensure that building projects comply with local building codes, zoning ordinances, and safety standards. In the context of off-grid living, obtaining the necessary building permits for your structures is essential for avoiding legal complications and potential fines.

When embarking on an off-grid building project, whether it's a cabin, a greenhouse, or a composting toilet, it's important to research and understand the specific building permit requirements in your area. While some jurisdictions may have lenient regulations for small-scale off-grid structures, others may impose stringent requirements, particularly in terms of structural integrity, sanitation, and environmental impact.

Consulting with local building authorities and zoning officials is a crucial first step in the permit acquisition process. They can provide valuable guidance on the specific permits needed for your project, as well as any zoning restrictions or building code requirements that may apply. Be prepared to submit detailed plans and specifications for your off-grid structures, including construction materials, site layout, and proposed utilities.

Navigating Zoning Regulations:

Zoning regulations govern how land can be used within a particular area, dictating permissible land uses, building heights, setbacks, and lot sizes. These regulations are designed to promote orderly development, protect property values, and ensure compatibility between neighbouring land uses. When planning off-grid structures, it's essential to understand and comply with the zoning regulations applicable to your property.

Off-grid living presents unique challenges in terms of zoning compliance, as many traditional zoning ordinances are tailored to accommodate conventional utility infrastructure and residential development patterns. As such, off-grid dwellings may encounter zoning restrictions related to utility connections, habitable floor area, and property setbacks.

To navigate zoning regulations effectively, conduct thorough research into the zoning ordinances governing your property. Identify the zoning classification assigned to your land and review the corresponding regulations pertaining to residential development, accessory structures, and alternative energy systems. In some cases, seeking variances or special use permits may be necessary to accommodate off-grid structures that deviate from standard zoning requirements.

Building permits and zoning regulations are essential considerations for anyone embarking on an off-grid living journey. By understanding and complying with these regulatory frameworks, you can ensure the legality, safety, and sustainability of your off-grid structures. Remember to consult with local authorities, research applicable regulations thoroughly, and seek professional guidance if needed. With careful planning and adherence to regulatory requirements, you can build a thriving off-grid homestead that is both harmonious with its surroundings and compliant with local laws.

Water Rights and Rainwater Harvesting Regulations

In the parched landscape of off-grid living, water is not just a precious resource—it's a lifeline, a source of sustenance, and a symbol of resilience in the face of adversity. But as our intrepid family soon discovered, navigating the murky waters of water rights and rainwater harvesting regulations can be a daunting task, fraught with legal complexities and bureaucratic hurdles.

Water rights, the legal entitlement to use water from a particular source, vary widely depending on location and jurisdiction. In some areas, water rights are tied to land ownership, while in others, they are allocated based on historical usage or seniority. Understanding your water rights is essential for off-grid living, as it dictates your ability to access and use water for drinking, irrigation, and other essential needs.

But perhaps the most pressing concern for off-grid dwellers is the thorny issue of rainwater harvesting regulations. In many regions, rainwater harvesting is subject to strict regulations and permitting requirements, designed to ensure the sustainable management of water resources and prevent over-extraction from natural sources.

For our family, navigating these regulations was a formidable challenge, requiring patience, perseverance, and a deep understanding of local laws and ordinances. They spent countless hours researching zoning regulations, studying water rights statutes, and consulting with legal experts to ensure compliance with the law while maximizing their water harvesting potential.

But despite the bureaucratic hurdles, our family remained undeterred in their quest for water independence. They embraced innovative solutions like rooftop rainwater harvesting systems, which capture and store rainwater for household use, reducing their reliance on dwindling groundwater reserves and municipal water supplies.

Through trial and error, our family learned valuable lessons about the importance of responsible water management and the need to work within the confines of the law to achieve their goals. They discovered that while rainwater harvesting regulations may pose challenges, they also present opportunities for creativity, innovation, and collaboration with local authorities and community stakeholders.

So, as you embark on your own off-grid journey, remember to familiarize yourself with water rights and rainwater harvesting regulations in your area. Arm yourself with knowledge, consult with local experts, and advocate for policies that promote sustainable water management and equitable access to this precious resource. For in the vast expanse of off-grid living, water rights are not just legal constructs—they're the foundation of a resilient, water-secure future for generations to come.

Animal Husbandry Regulations

In the idyllic realm of off-grid living, where the earthy scent of freshly tilled soil mingles with the gentle bleating of goats and the cheerful clucking of chickens, animal husbandry is a cornerstone of self-sufficiency. But as our intrepid family soon discovered, navigating the regulatory landscape of raising livestock in a remote wilderness is not without its challenges.

Animal husbandry regulations vary widely from region to region, encompassing everything from zoning restrictions to health and safety standards. Before embarking on your own off-grid homestead, it's crucial to familiarize yourself with the specific regulations governing the care and keeping of livestock in your area.

Zoning regulations are often the first hurdle aspiring off-grid homesteaders encounter when seeking to raise livestock. These regulations dictate the types and number of animals permitted on your property, as well as the location and size of their enclosures. Whether you're dreaming of raising a small flock of chickens or a herd of goats, zoning regulations will play a key role in determining the feasibility of your animal husbandry endeavours.

Health and safety standards are another important consideration when raising livestock off the grid. These regulations govern everything from the cleanliness of animal housing to the handling and disposal of animal waste. Failure to comply with these standards can not only endanger the health and well-being of your animals but also run afoul of local authorities, leading to fines and other penalties.

But navigating animal husbandry regulations isn't just about compliance—it's also about ensuring the health, safety, and welfare of your animals. From providing ample space and shelter to implementing sound husbandry practices, it's incumbent upon off-grid homesteaders to prioritize the well-being of their animal companions and adhere to the highest ethical standards of care.

For our intrepid family, navigating animal husbandry regulations was a journey marked by diligence, patience, and a deep respect for the natural world. By taking the time to familiarize themselves with the regulations governing livestock care in their area and implementing robust husbandry practices, they were able to raise healthy, happy animals while staying in compliance with local laws.

So, as you embark on your own off-grid homesteading journey, remember to do your homework, familiarize yourself with the regulations governing animal husbandry in your area, and prioritize the health and well-being of your animal companions above all else. With careful planning, sound judgment, and a commitment to ethical stewardship, you can raise livestock off the grid in harmony with both nature and the law.

CHAPTER 13
BUILDING A COMMUNITY: OFF-GRID LIVING IN GROUPS

In the vast wilderness of off-grid living, there exists a profound paradox—a simultaneous longing for solitude and a deep-seated desire for community. For our intrepid family, this dichotomy became increasingly apparent as they navigated the trials and triumphs of self-sufficiency, yearning for the camaraderie of like-minded souls even as they relished the solitude of their remote homestead.

As they gazed out upon the untamed beauty of their surroundings, our family couldn't help but feel a sense of longing—a yearning for connection, for companionship, for the shared joys and sorrows of communal living. And so, they embarked on a quest to build a community—a vibrant tapestry of individuals bound together by a shared vision of off-grid resilience, sustainability, and solidarity.

But building a community in the wilderness was no easy task. It required courage, determination, and a willingness to step outside the comfort zone of individualism and embrace the transformative power of collective action. Yet, with each passing day, our family discovered that the rewards far outweighed the challenges—a sense of belonging, a network of support, and a shared sense of purpose that transcended the boundaries of individual ambition.

As they gathered around the flickering warmth of a communal fire, our family found solace in the shared stories and laughter of their newfound tribe. Together, they shared the burdens and blessings of off-grid living, pooling their resources, skills, and knowledge to create a thriving ecosystem of mutual aid and cooperation.

But community living wasn't just about sharing resources—it was about sharing dreams, aspirations, and a vision for a better world. From communal gardens and shared livestock to cooperative energy projects and collective decision-making processes, our family discovered that the true essence of off-grid community living lay not just in self-sufficiency, but in interdependence—a recognition of the interconnectedness of all life and the power of solidarity in the face of adversity.

And so, as they looked out upon the horizon, our family saw not just a homestead, but a vibrant community—a tapestry of diverse individuals united by a shared commitment to living lightly on the land and nurturing the delicate balance of nature. Together, they forged a new path—a path towards a future where off-grid living wasn't just a solitary pursuit, but a collective endeavour—a journey of shared joys, shared challenges, and shared triumphs.

As you embark on your own journey towards off-grid living, remember that you are not alone. Whether you're building a community from the ground up or joining an existing tribe, know that there are others out there who share your vision, your values, and your dreams. Together, we can create a world where off-grid living isn't just a lifestyle choice, but a vibrant community—a tapestry of resilience, solidarity, and hope for generations to come.

Benefits and Challenges of Community Living

In the vast landscape of off-grid living, the concept of community is as essential as sunlight and soil. For our intrepid family, the decision to embrace community living was not just a practical choice but a deeply profound one—a commitment to forging bonds of solidarity, sharing resources, and cultivating a spirit of cooperation in the face of adversity.

Imagine their delight as they stumbled upon a thriving off-grid community nestled in the heart of the wilderness—a patchwork of cabins and gardens, each one a testament to the resilience and ingenuity of its inhabitants. Here, amidst the rustling leaves and chirping birds, they found a sense of belonging, a kinship that transcended mere friendship and bloomed into something akin to family.

But the benefits of community living extended far beyond mere companionship. In this close-knit enclave, our family discovered a wealth of shared knowledge and expertise, a collective wisdom born of years of trial and error. From seasoned gardeners to master craftsmen, each member of the community brought their own unique skills to the table, enriching the collective experience and ensuring that no challenge went unanswered.

Together, they tackled the monumental task of building a sustainable future, pooling their resources and labour to create a thriving oasis of self-sufficiency in the wilderness. From communal gardens bursting with bounty to shared workshops buzzing with creativity, their community became a beacon of hope in a world grappling with environmental crisis—a living testament to the power of cooperation and solidarity in the face of adversity.

Yet, for all its benefits, community living was not without its challenges. In the close quarters of communal living, conflicts inevitably arose, tensions simmered, and egos clashed. From disagreements over resource allocation to disputes over property boundaries, our family soon learned that living in harmony with others required patience, empathy, and a willingness to compromise.

But it was in the crucible of conflict that they discovered the true strength of their community—a resilience born of mutual respect, understanding, and a shared commitment to the greater good. Through open dialogue, respectful communication, and a steadfast dedication to the principles of cooperation, they navigated the rocky shoals of community living with grace and humility, emerging stronger and more united than ever before.

In the end, the benefits of community living far outweighed the challenges. For our family, the sense of belonging, the shared camaraderie, and the collective sense of purpose were more than worth the occasional disagreement or inconvenience. In their community, they found not just a place to live, but a home—a sanctuary of solidarity, resilience, and hope in a world desperately in need of all three.

Sharing Resources and Expertise

In the rugged terrain of off-grid living, community is not just a luxury—it's a lifeline. As our intrepid family embarked on their journey towards self-sufficiency, they quickly realized the power of sharing resources and expertise, forging bonds with like-minded individuals who shared their passion for sustainable living and off-grid resilience.

Imagine their delight as they stumbled upon a thriving community of fellow homesteaders, each one eager to share their knowledge, wisdom, and hard-won lessons from the field. From seed swaps to barn raisings, these communal gatherings became the beating heart of their off-grid existence, a place where friendships blossomed, ideas flourished, and dreams took root in the fertile soil of collective effort.

But the benefits of community extended far beyond the exchange of practical skills and resources. In the often-isolating world of off-grid living, these communal gatherings provided much-needed camaraderie and support, offering a shoulder to lean on during tough times and a chorus of cheers during moments of triumph. Whether sharing stories around the campfire or lending a helping hand in times of need, their fellow homesteaders became not just neighbours, but cherished friends and allies in the quest for self-sufficiency.

Yet, perhaps the greatest gift of community was the opportunity to pool their collective resources and expertise, tackling challenges that would be insurmountable for any one individual alone. From constructing communal root cellars to organizing bulk food purchases, these collaborative efforts allowed our family to achieve feats of self-sufficiency that would have been impossible on their own, transforming their off-grid existence from a solitary struggle into a shared adventure.

So, as you embark on your own journey towards self-sufficiency, remember the power of community—the power of sharing resources and expertise, of forging bonds with like-minded individuals who share your passion for sustainable living and off-grid resilience. Whether online or in person, seek out your fellow homesteaders, your fellow dreamers, and your fellow travellers on the road to self-sufficiency. For in their wisdom, their knowledge, and their unwavering support, you'll find the strength and inspiration to thrive in the wild embrace of off-grid living.

Barter Systems and Local Co-ops

In the vast expanse of off-grid living, there exists a profound truth: no one can thrive in isolation. As our intrepid family embarked on their journey towards self-sufficiency, they quickly discovered the importance of building a strong and resilient community—a network of like-minded individuals bound together by a shared vision of a more sustainable and interconnected world.

At the heart of this community lay the concept of barter systems and local co-ops, two pillars of solidarity and mutual support that helped our family navigate the challenges of off-grid living with grace and resilience.

Barter Systems:

Picture this: Our family finds themselves with an abundance of fresh produce from their bountiful garden, yet lacking in essential tools for their homestead. Enter the barter system—a time-honoured tradition that transcends the confines of currency, allowing individuals to exchange goods and services directly with one another.

Through the barter system, our family was able to trade their surplus vegetables for the tools they needed, forging connections with their neighbours and fostering a spirit of reciprocity and goodwill. It was a symbiotic relationship built on trust and mutual benefit, where everyone had something to offer and something to gain.

But the barter system was more than just a means of acquiring material goods—it was a catalyst for community building, a bridge that connected individuals from all walks of life in a shared pursuit of self-sufficiency and resilience. In the bustling marketplace of the off-grid world, our family found not just the tools they needed, but also the camaraderie and support of their fellow homesteaders, transforming their isolated cabin in the woods into a vibrant hub of connection and collaboration.

Local Co-ops:

Yet, the spirit of cooperation didn't end at the barter table—it extended into the realm of local co-ops, grassroots organizations dedicated to pooling resources, sharing knowledge, and promoting sustainable living practices within the community.

Through their local co-op, our family gained access to a wealth of resources and expertise, from bulk purchasing opportunities for essential supplies to workshops and skill-sharing events that empowered them to deepen their understanding of off-grid living.

But perhaps most importantly, the local co-op served as a hub of connection and belonging—a place where our family could come together with their neighbours to celebrate their successes, support one another through their challenges, and dream of a brighter, more resilient future.

In the end, it was the combined power of barter systems and local co-ops that helped our family not just survive, but thrive in the wild embrace of off-grid living. Through these pillars of solidarity and mutual support, they discovered that true self-sufficiency isn't just about individual resilience—it's about building a community that thrives together, united in a shared vision of a more sustainable and interconnected world.

Part 6
Appendix

Resource Guide - Books, Websites, and Organizations

In the vast landscape of off-grid living, knowledge is not just power—it's survival. As our intrepid family embarked on their journey towards self-sufficiency, they quickly realized the importance of arming themselves with the right resources. From trusted books and informative websites to invaluable organizations dedicated to sustainable living, they sought out a wealth of information and support to guide them through the trials and triumphs of off-grid living.

Books:

1. ***The Encyclopedia of Country Living* by Carla Emery:** This timeless classic is a comprehensive guide to homesteading and self-sufficiency, covering everything from gardening and food preservation to animal husbandry and renewable energy.
2. ***The Backyard Homestead* by Carleen Madigan:** Packed with practical tips and innovative ideas, this book offers a wealth of advice on maximizing your space and resources to create a self-sufficient homestead right in your own backyard.
3. ***The Off-Grid Living Handbook* by Teri Page:** Drawing on her own experiences of off-grid living, Teri Page offers invaluable insights and practical advice on everything from building a solar power system to growing your own food and preserving the harvest.
4. ***The Self-Sufficient Life and How to Live It* by John Seymour:** A classic guide to self-sufficiency, this book covers all aspects of off-grid living, from growing your own food to building your own shelter and generating your own power.
5. ***The Solar Electricity Handbook* by Michael Boxwell:** For those looking to harness the power of the sun, this comprehensive guide offers practical advice on designing, installing, and maintaining a solar power system for your off-grid home.

Websites:

1. **Mother Earth News**: A treasure trove of articles, guides, and how-to videos on all aspects of sustainable living, from organic gardening to renewable energy and beyond.
2. **Premise**: An online community dedicated to permaculture and regenerative living, offering forums, courses, and resources for those looking to create resilient, self-sustaining ecosystems.
3. **Off-Grid World**: A hub of information and inspiration for off-grid enthusiasts, featuring articles, product reviews, and real-life stories from people living off the grid around the world.
4. **Backwoods Home Magazine**: A source of practical advice and homesteading wisdom, with articles on everything from building your own cabin to raising livestock and preserving food.
5. **Renewable Energy World**: A leading source of news and information on renewable energy technologies, offering insights into the latest developments in solar, wind, hydro, and more.

Organizations:

1. **The Permaculture Research Institute**: Dedicated to promoting sustainable agriculture and regenerative design, this organization offers courses, workshops, and resources for aspiring Perma culturists around the world.
2. **The American Solar Energy Society (ASES)**: A nonprofit organization dedicated to advancing the use of solar energy for the benefit of humanity, with resources for homeowners, businesses, and policymakers interested in solar power.
3. **The Sustainable Living Association**: Committed to fostering sustainable communities and lifestyles, this organization offers educational programs, events, and resources to help people live more sustainably.
4. **The Land Institute**: Working towards a future where agriculture is in harmony with nature, this organization conducts research and advocates for perennial polyculture farming systems that mimic the structure and function of natural ecosystems.
5. **The Off-Grid Living Network**: A community of like-minded individuals and families living off the grid, offering support, advice, and resources for those seeking to embrace a simpler, more sustainable way of life.

With these resources at their fingertips, our intrepid family embarked on their journey towards self-sufficiency with confidence and determination. And as you embark on your own off-grid adventure, may these books, websites, and organizations serve as guiding lights, illuminating the path to a brighter, more sustainable future for us all.

Glossary of Terms

1. **Amp-hours:** A unit of measurement for electrical energy storage capacity, commonly used in batteries.
2. **Biofuel:** Renewable fuel derived from organic materials, such as plant matter or animal waste.
3. **Composting toilet:** A toilet that uses the natural process of decomposition to break down human waste into compost.
4. **Greywater:** Wastewater generated from household activities like bathing and laundry, excluding toilet waste.
5. **Homesteading:** A lifestyle of self-sufficiency characterized by subsistence agriculture, food preservation, and small-scale livestock raising.
6. **Micro hydropower:** Small-scale hydroelectric power generation using the flow of water from a small stream or river.
7. **Off-grid living:** A lifestyle independent of public utilities, such as electricity, water, and sewage systems.
8. **Passive heating and cooling:** Design principles and techniques that utilize natural elements, such as sunlight and airflow, to regulate indoor temperature.
9. **Permaculture:** A design system that integrates sustainable agriculture, ecological principles, and community development to create harmonious human habitats.
10. **Solar dehydrator:** A device that uses solar energy to remove moisture from food, preserving it for long-term storage.
11. **Solar oven:** A device that uses solar energy to cook food without the need for conventional fuels.
12. **Wind turbine:** A device that converts wind energy into electrical power through the rotation of blades connected to a generator.
13. **Zoning regulations:** Local government ordinances that dictate land use, building construction, and development standards within designated zones or districts.
14. **Battery bank:** A collection of interconnected batteries used to store electrical energy generated from renewable sources, such as solar panels or wind turbines.
15. **Biogas digester:** A system that breaks down organic waste, such as animal manure or food scraps, to produce biogas, a renewable fuel composed primarily of methane.
16. **Canning:** A food preservation method that involves sealing food in jars or cans and heating them to kill bacteria, ensuring long-term storage stability.
17. **Distillation:** A process of purifying water by boiling it and then condensing the steam back into liquid form, effectively removing impurities.

18. **Emergency preparedness:** The process of planning and preparing for potential emergencies or disasters, including natural disasters, power outages, or other unforeseen events.
19. **First aid:** Immediate medical assistance provided to individuals who have been injured or are experiencing sudden illness, typically administered until professional medical help arrives.
20. **Homesteading skills:** Practical abilities related to self-sufficient living, including gardening, animal husbandry, food preservation, and basic carpentry.
21. **Off-grid communication:** Methods and technologies used to communicate in remote or isolated areas without access to traditional telecommunications infrastructure.
22. **Root cellar:** An underground storage space used to store fruits, vegetables, and other perishable food items at cool temperatures, preserving them for extended periods.
23. **Seed selection:** The process of choosing and sourcing seeds for planting based on factors such as climate, soil conditions, and desired crop varieties.
24. **Sustainable waste management:** Practices and strategies for minimizing waste generation, reusing materials, and recycling or composting organic waste to reduce environmental impact.
25. **Water purification:** The process of removing contaminants and impurities from water to make it safe for drinking or other uses, often involving filtration, disinfection, or chemical treatment methods.
26. **Ampere (A):** The unit of measurement for electric current, representing the flow of electrical charge through a conductor.
27. **Biomass:** Organic matter derived from living or recently living organisms, often used as a renewable energy source through processes like combustion or fermentation.
28. **Carbon footprint:** The total amount of greenhouse gases, particularly carbon dioxide, emitted directly or indirectly by an individual, organization, or activity.
29. **Homestead Act:** Legislation passed in the United States in 1862 that provided settlers with land grants in exchange for cultivating and improving the land.
30. **Inverter:** A device that converts direct current (DC) electricity from sources like solar panels or batteries into alternating current (AC) electricity for use in household appliances.
31. **Passive solar design:** Architectural techniques that maximize the use of sunlight for heating, lighting, and ventilation in buildings without the need for mechanical systems.

32. **Percolation:** The process by which water seeps through soil layers, often used to assess the suitability of soil for septic systems or rainwater infiltration.
33. **Photovoltaic (PV) panel:** A device that converts sunlight directly into electricity using semiconductor materials, commonly used in solar power systems.
34. **Rainwater harvesting:** The collection and storage of rainwater for various uses, such as irrigation, toilet flushing, or household consumption.
35. **Renewable energy:** Energy derived from naturally replenished sources, such as sunlight, wind, or flowing water, that can be used indefinitely without depletion.
36. **Solar insolation:** The amount of solar radiation received per unit area at a given location, typically measured in kilowatt-hours per square meter per day.
37. **Sustainability:** The practice of meeting present needs without compromising the ability of future generations to meet their own needs, often encompassing environmental, social, and economic dimensions.
38. **Thermal mass:** The ability of a material to absorb and store heat energy, often used in passive solar heating systems to regulate indoor temperatures.
39. **Vermicomposting:** A composting method that uses earthworms to break down organic waste into nutrient-rich compost, suitable for use in gardening and soil improvement.
40. **Water rights:** Legal entitlements to use water resources, typically governed by laws and regulations that allocate water for various purposes, such as irrigation, industry, or domestic use.

Bonus
Additional Resources and Worksheets

In addition to the wealth of information provided in this book, we've curated a bonus section packed with valuable resources and practical worksheets to support you on your journey towards self-sufficiency and off-grid living. From helpful websites and recommended reading to actionable worksheets for planning and implementation, these resources are designed to empower you to thrive in the wild embrace of self-reliance.

1. Recommended Reading

Explore these recommended books and websites for further insights and inspiration on off-grid living, sustainable practices, and homesteading skills:

- "The Backyard Homestead" by Carleen Madigan
- "The Self-Sufficient Life and How to Live It" by John Seymour
- "Mother Earth News" (Website)
- "Off Grid World" (Website)
- "Permaculture Principles" (Website)

These resources offer a wealth of knowledge and practical advice from experts in the field, providing invaluable guidance for your off-grid journey.

2. Practical Worksheets

Put your newfound knowledge into action with these practical worksheets designed to help you plan, implement, and track your off-grid projects:

- Solar System Sizing Worksheet: Calculate your energy needs and design a solar power system tailored to your requirements.
- Garden Planning Worksheet: Plan your off-grid garden layout, select suitable crops, and schedule planting dates for optimal yield.
- Water Collection System Design Worksheet: Design and map out your rainwater harvesting system, including collection methods, storage capacity, and filtration options.
- Emergency Preparedness Checklist: Prepare for unexpected events with this comprehensive checklist covering essential items and tasks for emergency readiness.
- Monthly Sustainability Tracker: Monitor your progress towards self-sufficiency with this monthly tracker, recording key metrics such as energy consumption, food production, and waste reduction.

These worksheets serve as practical tools to guide you through the implementation of your off-grid projects, empowering you to take tangible steps towards a more sustainable lifestyle.

Unlock the full potential of your off-grid lifestyle with these bonus resources and worksheets, designed to complement the knowledge and skills imparted in this book. Embrace the journey towards self-sufficiency with confidence, creativity, and a sense of adventure, knowing that you have the tools and support you need to thrive in the wilderness of off-grid living.

www.ingramcontent.com/pod-product-compliance
Lightning Source LLC
Chambersburg PA
CBHW082210220526
45470CB00010B/3114